T0093763

The ABC's of Science

Giuseppe Mussardo

The ABC's of Science

 Springer

Giuseppe Mussardo
SISSA
Trieste, Italy

ISBN 978-3-030-55168-1   ISBN 978-3-030-55169-8   (eBook)
https://doi.org/10.1007/978-3-030-55169-8

© Springer Nature Switzerland AG 2020
Italian language edition: L'alfabeto della scienza. Da Abel a Zero assoluto 26 storie di ordinaria genialità
by Giuseppe Mussardo, Edizioni Dedalo 2020.
This work is subject to copyright. All rights are reserved by the Publisher, whether the whole or part
of the material is concerned, specifically the rights of translation, reprinting, reuse of illustrations,
recitation, broadcasting, reproduction on microfilms or in any other physical way, and transmission
or information storage and retrieval, electronic adaptation, computer software, or by similar or dissimilar
methodology now known or hereafter developed.
The use of general descriptive names, registered names, trademarks, service marks, etc. in this
publication does not imply, even in the absence of a specific statement, that such names are exempt from
the relevant protective laws and regulations and therefore free for general use.
The publisher, the authors and the editors are safe to assume that the advice and information in this
book are believed to be true and accurate at the date of publication. Neither the publisher nor the
authors or the editors give a warranty, expressed or implied, with respect to the material contained
herein or for any errors or omissions that may have been made. The publisher remains neutral with regard
to jurisdictional claims in published maps and institutional affiliations.

Cover picture and the 26 drawings in the book: © Debora Gregorio

This Springer imprint is published by the registered company Springer Nature Switzerland AG
The registered company address is: Gewerbestrasse 11, 6330 Cham, Switzerland

The original version of the Book Frontmatter has been revised.
Cover picture and the 26 drawings in the book: © Debora Gregorio
Italian language edition: L'alfabeto della scienza. Da Abel a Zero assoluto 26 storie
di ordinaria genialità by Giuseppe Mussardo, Edizioni Dedalo 2020.

Preface

There is a passage in Isaac Asimov's *Guide to Science for Modern Man* that has always stayed with me, ever since I read it many years ago: It is one of those damn perfect phrases that push you to look beyond the usual horizons and that made Asimov the great writer that he was:

> Nobody can really feel comfortable in the modern world and evaluate the nature of his problems—and their possible solutions—without having an exact idea of what science is doing. Furthermore, initiation into the wonderful world of science is a source of great aesthetic satisfaction, inspiration for young people, fulfillment of the desire to know and a deeper appreciation of the admirable potential and capacity of the human mind.

I would like to add that the most authentic and genuine spirit of science emerges not only from its direct study—the effort to understand the thousands of theorems that form the architecture of mathematics or those five or six fundamental physical ideas that shape our world view—but also directly from the history of its protagonists, from their voices, from that combination of characters, scenes, and stories that form the complexity and beauty of the scientific enterprise, made of laboratory papers, notebooks full of chunks of formulae, ideas written on the margins of books, peaceful discussions, and fierce rivalries.

Obviously, there is no single scientific personality: every scientist, on the contrary, is genuinely wacky, unpredictable, exalted, and bizarre in his or her own way. As in art, painting or music, so in science can we find those who are driven by instinct, some by passion, some by caution, some by precision, some by love of numbers, and some by the beauty of theories. To these thousands of facets of the human soul, and to the rigor and sharpness of intelligence, we must add curiosity—the real wellspring of all science—and also the thrill of the new challenges to one's knowledge, the litmus test of the true scientist.

The extraordinary thing is that, alongside the justifiably most famous figures—Albert Einstein, Enrico Fermi, Galileo Galilei, etc.—there is an incredible crowd of other scientists, who, although unknown to most, have taken great steps forward in our knowledge with their incurable curiosity. Some of them have also posed great questions, even some of a moral nature: as a matter of fact, separating science from the dizzying flow of history, with all of its anxieties, its dramas and its paradoxes, is an unrealistic exercise, though perhaps fascinating from a philosophical point of view.

Here, in this book, the reader can find twenty-six of these characters, a number certainly too small to fully convey the beauty of the fascinating journey made over the centuries in search of the truth, but hopefully large enough to reveal the more authentic and polymorphic spirit of the world of science. We will talk about great topics and great scientific ideas—the arrow of time, the birth and death of stars, the absolute zero of temperatures, the beauty of mathematics, and the transmutation of the elements, just to name a few—but we shall also talk, above all, about those people, men and women, who have revealed the mysteries of Nature to us: In fact, there seems to be a graceful and mysterious duality that has always linked people to the world of ideas, and vice versa.

Trieste, Italy   Giuseppe Mussardo

Contents

Abel. An Elliptic Thriller

Königsberg, a city on the eastern border of Prussia, was one of the most flourishing ports in the Baltic Sea in the nineteenth century. Within the circles of philosophers, it was known as "the city of Immanuel Kant": the great German thinker was born and had lived there, and he had always led an extremely regular and habitual life; the locals used to set their clocks when they saw him passing by. Contrastingly, within mathematical circles, Königsberg was instead known as "the city of the seven bridges": placed at the confluence of two rivers, the town was divided into four parts, connected to each other by seven bridges. According to a legend, there was a bet about the possibility of finding a route that, starting from one of the four areas of the city and crossing each bridge only once, would allow one ultimately to be able to return to the starting point. But, no matter how they tried, nobody was ever able to come up with a solution: the problem truly seemed to be impossible...

The Prussian style of the city led to a rigid, punctual and meticulous way of life, with people being careful about money and paying attention to the details of things: for example, on October 4, 1827, the librarian of the University of Königsberg sent a letter to August Leopold Crelle, a Berlin publisher, complaining that the latest issue of the scientific journal that he edited had been sent to the University by courier and not by regular mail, causing an additional cost of a thaler for the library coffers! This was, of course, a purely Prussian zeal, but it is perhaps thanks to the diligence of that librarian that we can shed light on one of the most controversial and compelling scientific competitions in Mathematics: the competition between Niels Henrik Abel and Carl Gustav Jacobi, both in their early twenties. The thorny question of the priority of their discoveries in a very fascinating field of mathematics—that of elliptic functions—seems almost a spy story, so romantic and complicated that time has helped to bury it or, at least, admirably hide it within the intricate chronology of events or between the lines of diaries or epistolaries of the period. Indeed, a tragic story.

© Springer Nature Switzerland AG 2020
G. Mussardo, *The ABC's of Science*,
https://doi.org/10.1007/978-3-030-55169-8_1

On the Streets of Paris

According to Honoré de Balzac, "Paris is in truth an ocean: you can plumb it but you'll never know its depths." Niels Abel had arrived there for the first time in July 1826, the hottest month of the year and in the middle of the holidays. The universities were closed and most of the scholars and professors were enjoying the fresh air in the country houses outside of the city. It was the ideal time to calmly come to grips with his many projects and to improve his French, so that he would be ready to discuss any topic with all of the wise and famous men who filled the city: Laplace, Cauchy, Legendre, Fourier... He was twenty-four years old and had enormous expectations for the meeting with all of those great mathematicians, all of whose original works he had avidly studied when he was in high school and at the University of Christiania in Norway: "Learn from the masters and not from the disciples," this was his principle.

Abel was the son of a Protestant pastor and had grown up in the village of Finnøy, surrounded by frozen lakes and a countryside almost always blanketed with snow: his face was as beautiful as his mother's, with large blue eyes and a tuft of hair that fell across his forehead. His childhood had been marked by poverty and melancholy, dinners based on potatoes or herring, but also by a boundless passion for mathematics, which he had approached at a very young age thanks to the suggestions and stimuli of Bernt Michael Holmboe, his high school professor: he had read the works of Newton, those of Euler, Gauss's *Disquisitiones Arithmeticae*—the famous book about number theory—and many other texts.

Poverty was also soon followed by tragedy: his father died when he was just eighteen years old, and he was forced to take care of his mother and his six brothers, a task that absorbed a lot of his energy, but which he always carried on with a smile that illuminated his beautiful face. During his few free moments, he was absorbed by a very interesting mathematical problem: finding the general solution formula to a fifth-degree algebraic equation, a problem that had fascinated many mathematicians since the Renaissance. Although he did not have much time to think about it, he was nevertheless able to come to an amazing conclusion, namely, that finding a general solution to the fifth degree equation might be an impossible task! It was an exceptional result, and later became one of history's most famous mathematical theorems. Just as the great Euler, at the end of the eighteenth century, had demonstrated the impossibility of solving the problem of the Königsberg bridges, so Abel had come to the conclusion that even the world of algebraic equations can be ruled by some strange impossibilities: this discovery only increased his admiration for mathematics, a discipline so powerful that it even manages to identify its own limits.

This result had given him some notoriety at home and had earned him unconditional support from all of the professors at the University of Christiania, where he had been enrolled since 1821, and of Holmboe, who, in the meantime, had become his great friend. It was, in fact, thanks to their efforts that he had won a grant from the Ministry of Education—a modest one, to be honest—to tour the European continent in order to meet with the best German and French mathematicians of the time. Abel

hoped very much that he would be welcomed, as well as entertaining the possibility that he might get a chair in one of Europe's prestigious universities, so that he could solve his serious economic problems once and for all. However, he had not taken into account the obtuse indifference of many great mathematicians towards a young unknown man from the distant Nordic countries.

Once he had left Norway, he initially spent several months in Germany, where he made a conscious choice not to meet Gauss, who had made unflattering comments about his result on the algebraic fifth degree equations. Gauss was also known as the prince of mathematicians, but, on that occasion, he had certainly not shown a noble etiquette: "That is a problem that has been open for a hundred years: if it has not been solved by the best mathematicians, how can you believe that it could have been solved by a young Norwegian man? It must certainly be a hoax"; these were more or less his words. For Abel, they had been a source of real disappointment. Therefore, instead of Göttingen, he decided to go to Berlin, where he immediately attracted the sympathies of August Leopold Crelle, an engineer with an immense passion for mathematics and a particular intuition for discovering new talents. It was fortunate for both. Years later, Crelle still remembered how they met: "One day a boy with a smart appearance knocked on my door, very embarrassed. I thought he was there for the entrance exam to the School of Commerce for which I was then responsible, and therefore I began to list all the subjects he had to know, the details of the programmes and similar things. Looking at me straight in the eye, the boy finally opened his mouth and, in bad German, said: "No exam, only math!" And, indeed, we immediately started to discuss math!"

Right around the time of their meeting, to promote mathematics—a discipline that he truly loved—Crelle had founded the "Journal für die reine und angewandte Mathematik" (Journal of Pure and Applied Mathematics), a quarterly journal that would go on to play a very important role in future decades, indeed becoming one of the best mathematics journals in Europe. The policy of the journal was strict: review papers or papers addressing uninteresting problems would have no chance of being published. With Abel, Crelle immediately understood that he had stumbled upon a genius: he loved hearing Abel talk about mathematics, following his arguments or his geometric intuition on many problems. He was therefore delighted to publish seven papers by Abel, the Norwegian prodigy, in the first three issues of his "Journal," including the paper concerning the impossibility of solving algebraic equations of a degree higher than the fifth through the use of radicals.

After his stay in Germany, Abel decided to move to Paris. Upon arrival, he took lodging in the Faubourg Saint-Germain and then waited for the end of summer. In those calm months, he completed a paper that was destined to change the history of mathematics forever: it concerned the study of "a special class of transcendent functions," a subject on which many other mathematicians before him had broken their heads, primarily Legendre, but also Cauchy and Gauss. He therefore eagerly wanted to illustrate his results to all of those great mathematicians, to discuss with them the new horizons that these discoveries opened up, to hear their comments...

Box. Elliptic Functions

Elliptic functions are one of the masterpieces of nineteenth-century mathematics. Giulio Carlo di Fagnano, Leonhard Euler, Carl Gauss and Adrien-Marie Legendre were among the main architects of the first developments in this field. Niels Abel and Carl Jacobi, for their part, were rather the main characters in the compelling scientific competition that lead to the definitive formulation of the theory. Significant contributions also came from Bernhard Riemann, Karl Weierstrass and Charles Hermite. A particular role was played by Carl Gauss, who was, in fact, the only one of his time, with Legendre, to fully understand the importance of the topic, since it had been the subject of his study for more than thirty years. However, all of his discoveries in this field were entrusted exclusively to the pages of his diaries, although some allusions can be found in his work Disquisitiones Arithmeticae of 1801.

Aside from the events involving Abel and Jacobi, the whole topic of elliptical functions constitutes a rather curious chapter in the history of mathematics. The sequence of discoveries proceeded in an almost circular way, taking its first moves from geometry (with the calculation of the length of particular algebraic curves) only to later move into the field of pure analysis (with the systematic examination of particular integral expressions), and then ultimately returning to the field of geometry, reaching its apotheosis with an extremely simple and elegant idea, that of the Riemann surfaces.

Thanks to the fundamental idea of inversion, carried out by Abel and Jacobi, it was finally understood that, in the complex plane, the elliptic functions are doubly periodic, and their periodicity cells provide, in particular, a tessellation of the plane, just like floor tiles. If Abel's greatest work is his memory published in Crelle's "Journal," for Jacobi, it is his treatise 'Fundamenta nova theoriae functionum ellipticarum,' written in 1829, when he was just twenty-five years old, a book destined to become, since its publication, the reference text par excellence of this discipline. So, it is wrong to believe that, if Abel was, as Felix Klein put it, the "Mozart of mathematics," Jacobi was the perfidious Salieri!

Unfortunately, none of this happened: everyone proved to be very civil, indeed, but no more than that, and if they argued with him, it was only to talk about *their* own work, while they never showed any interest in what Abel had to say. Cauchy exuded confidence, pedantry, and indifference. This was unbearable for the young Norwegian, who manifested all of his discouragement in a letter to his dear friend Holmboe, dated October 24, 1826:

Actually, this, which is the most chaotic and noisy capital of the continent, is only a desert for me... Legendre is extremely polite, but very old, Cauchy simply crazy. What he does is excellent, yet really messy. He does pure mathematics, while Fourier, Poisson, Ampère are busy all the time studying magnetism and other fields of physics. Laplace is a sprightly old man, who hasn't written for a long time... There is a great difficulty in getting in touch with them: everyone wants to teach and nobody wants to learn. Absolute selfishness reigns everywhere. I have just finished a long memory on a special class of transcendent functions that I would like to present at the Academy of Sciences. I showed it to Cauchy, who however did not give it neither a look. I think it's important, you don't know how much I'd like to know what the Academy thinks about it.

The "transcendent functions" that Abel was writing about were the elliptical functions, a subject in regard to which, however, a very dangerous competitor, Carl Gustav Jacobi, had appeared on the horizon.

An Enfant Prodige

In the early nineteenth century, Potsdam was a prosperous village of six thousand souls in a beautiful location in the countryside near Berlin. The Imperial Court had transformed the entire area around that spot into a kind of paradise, a place to relax and leave all of the worries of the capital behind. Thus, the creation of wonderful tree-lined avenues, the magnificent and well-kept gardens of Sanssouci and, together with these, the fountains and the ponds, the colonnades and the Orangerie, the botanical garden and the greenhouses, which made the small village of Potsdam the "Versailles of Prussia." A number of wealthy people and families of the middle or upper middle class—lawyers, doctors, teachers—attracted to the social possibilities offered by that environment and its lifestyle habits were soon added to the Court officials and aristocratic families of the earlier times. It was a life marked by the alternation of the seasons and those amusements—horseback riding, parties, visits from relatives— that were sometimes almost a nightmare for their guests, forced to feign maximum interest in quince jams or glasses of bitter liquor: these were the innocent privileges of the upper classes who had chosen the countryside as their place of residence.

It was in this peaceful Potsdam atmosphere that Carl Gustav Jacobi was born in 1804, the second son of a wealthy Jewish family: his father was a banker, close to the Court, while his mother came from a similarly wealthy local family. Young Carl grew up in a comfortable and stimulating environment. A maternal uncle took charge of his initial education and introduced him to both classics and elementary mathematics. Physically frail, despite possessing a quite large head, with a high and spacious forehead crowned by a mountain of dark hair, above curious eyes and a witty smile on his lips, Jacobi had the exact appearance of a child prodigy: and, as a matter of fact, at twelve years old, he passed the entrance exam to the Gymnasium in Potsdam, displaying a lively talent and exceptional early development.

He always had top marks in Latin, Greek and history. The discipline in which he excelled, however, was mathematics, for which he nurtured a real transport, so much so that, at a very early age, he indulged in reading the texts of Leonhard Euler and other great mathematicians. Once he had enrolled at the University, he decided to follow his passion, a choice that became a sort of consecration: he proved to be a champion at performing any type of calculation and, at the same time, one of the most tireless mathematicians that the history of this discipline remembers. To a friend, who was worried about his health for all of the energy that he dedicated to mathematics, Jacobi replied: "I am sure it will affect my health, so what? Only cabbages have no nerves or worries. And what do they get at the end from their state of perfect well-being?"

After earning his doctorate, when he was just twenty-one years old, Carl Jacobi obtained a teaching qualification, and he proved to have an innate didactic talent: his lessons stood out for their clarity and vivacity, finding captivating explanations for even the most difficult topics. He used to illustrate theorems with examples and thorough discussions; no formulae held any secrets for him. These exceptional qualities ended up attracting the attention of the university authorities, and thus opened the door to a successful academic career, with the offer of a professorship at the University of Königsberg.

When he arrived there in 1826, adjusting to the new academic environment was not easy, given his strong temperament, his marked careerism and the ease with which he succumbed to making sarcastic comments. However, he did not care much about the hostility that surrounded him, completely immersing himself in the study of the rising theory of elliptic functions: in a very short time, he obtained very relevant results, unaware of Abel's existence and what the Norwegian mathematician had already achieved.

However, while poor Abel was grappling with the cold indifference of French mathematicians, Jacobi was instead able to capture the attention of Adrien-Marie Legendre and, thanks to the enthusiastic support of the latter, he managed to secure the position of associate professor at the University of Königsberg. A frequent correspondence started between the two: on one side, the old French mathematician, on the other side the young German professor. The letters that they exchanged are unique documents within the history of mathematics, particularly useful for the reconstruction of the genesis of the theory of elliptic functions and the race that began between Niels Abel and Carl Jacobi.

Priority Issues

The subject of elliptical curves, the corresponding integrals and the elliptic functions that invert them is one of the jewels of nineteenth-century mathematics. In fact, it marvelously combines three of the greatest leitmotifs of mathematics: complex variable functions, geometry and arithmetic. Anyone interested in only one of these will find an inexhaustible mine of surprises in the elliptical functions, even though the extraordinary elegance of this latter subject only emerges from the synthesis of the three topics above.

Within the history of elliptic functions, for a long time, attention was catalyzed by the study of a certain type of integral, an activity that Adrien-Marie Legendre committed to with admirable constancy for almost forty years of his life: the simplest of these integrals it originated precisely from the calculation of the length of an ellipse and, for this reason, in his treatises, Legendre introduced the terminology of *elliptical integrals*. But when, in 1828, Legendre published his ponderous book on the subject, the volume was immediately outdated, surpassed by the new discoveries made in that same year by Abel and Jacobi! Despite his long years of study, Legendre had undeniably missed the opportunity to understand the profound geometric nature of elliptical integrals.

The turning point was when Niels Abel and Carl Jacobi made their dual appearance on the scene. Their competition was marked by tight and exciting rhythms, but also controversial aspects, as in other great contests in mathematics, such as those that, in the Renaissance, had set Niccolò Tartaglia in opposition to Gerolamo Cardano or, in the seventeenth century, Jakob to Johann Bernoulli (even though they were brothers!), or, during the eighteenth century, Isaac Newton to Gottfried von Leibniz. To give a full account of the exciting race between Abel and Jacobi, it would be necessary to give rise to an exciting historiographical investigation characterized by a rather unique clockwork mechanism, with the rhythm of time marked by the dates of upon which the manuscripts were written and those upon which they were published, the dates of dispatch of the magazine issues and those of their reception, the writing of various letters, with content that may at first sight appear insignificant, but, as a matter of fact, would be decisive for putting any piece of the puzzle in the right place.

First of all, what was the turning point in the subject? A deep understanding of elliptical integrals required the genesis of a fundamental idea, which could only mature with the progress made in the field of complex analysis in the first decades of the nineteenth century. The idea—simultaneously simple and ingenious—was to pay less attention to the original integral expressions, and rather focus on their *inverse functions*: entirely new perspectives were opened up in the field of transcendent functions through this idea, and new, very elegant results were quickly achieved in various branches of mathematics. Jacobi was so impressed by the power of such a simple concept that he often repeated that the key to success in mathematics lay in the motto: "The important thing is to invert."

Niels Abel had developed the idea of inversion behind elliptic functions in 1823 and had made it the key concept of the memory presented in 1827 to the Academy of Sciences of France, which was lost among the Cauchy papers (and found only in 1952, in the Moreniana Library in Florence!). But the same principle had formed the backbone of his masterful article 'Recherches sur les fonctions elliptiques,' which appeared in the second issue of Crelle's "Journal" of September 1827. The debut of this article could not have been more explicit: "In this memory, I propose to consider the inverse function, which is…"

Carl Jacobi, on the other hand, had come to this intuition during the summer of 1827, and had initially made it the subject of two notes, also published in September of that year in the magazine "Astronomische Nachrichten," dated, respectively, June 13 and August 2, 1827. In these articles, he included some special laws that he had discovered for the transformation of elliptical integrals, but without any proof of their validity. It was on the occasion of these discoveries that Jacobi wrote to Legendre for the first time, on August 5, 1827. After the introduction, in which he expressed his veneration for the elderly French mathematician, then seventy-five years old, Jacobi provided a more detailed explanation of his formulae, undoubtedly seeking Legendre's response. His expectations were not disappointed: the Frenchman was already aware of Jacobi's results, following a reading of the "Astronomische Nachrichten" booklet, but, thanks to the letter, he also had a sense of the extraordinary prospects that these results could open up in the field to which he had dedicated forty years of his life. His enthusiasm was so strong that he talked about it at the November session of the Academy of Sciences in Paris, and the echoes of this talk did not take long to arrive in Germany: the news about Jacobi's equation was taken up and reported by various German newspapers, thus further magnifying the fame of the very young German mathematician.

In a further letter by Jacobi, dated January 12, 1828, besides warm thanks, which were not, after all, surprising, there were some other statements that were surprising indeed: after a few paragraphs, Jacobi mentioned "results of the greatest importance on the subject of Elliptical Functions, which were achieved by a young scholar whom you perhaps know personally." He was clearly talking about Niels Abel. Reading further down, one easily discovers that the entire letter was nothing more than a detailed exposition of the fundamental ideas credited to Abel! But where could Jacobi have learned them from?!

To clarify this issue, attention must be paid to the publication of issue 127 of "Astronomische Nachrichten" of December 1827; Jacobi himself had presented explicit proof of his initial results on the elliptical functions earlier, in September, in the same magazine. The initial absence of proof in the first article had raised criticisms by both Gauss and Legendre, and therefore, not surprisingly, Jacobi was very concerned about dissipating the doubts of the two great mathematicians. The date of the new manuscript, the one containing the proof of his initial result, was November 18, 1827. The importance of this article, however, is not so much in the anticipated demonstration of its previous formulae, but rather that it also contains the explicit definitions of the inverse functions of elliptical integrals: exactly the same ideas, formulae and definitions previously introduced in the paper 'Recherches sur les fonctions elliptiques' by Abel!

So, in a nutshell, the field of elliptical integrals that, since the year 1751, had achieved little progress, and only of a technical nature at that, had, within a period of less than two months, borne witness to a fundamental turning point thanks to two admirably similar works that both appeared at once. A circumstance that was, to say the least, curious, at least singular, but certainly strange.

This is precisely why the letter of complaint from the Königsberg library, written on October 4, 1827, helps us. Indeed, it allows us to say that the issue of the Crelle "Journal" of September 1827—the one with Abel's fundamental article on the inversion properties of elliptic functions—was already at the Königsberg library at the date of the letter, and therefore available for consultation by university members, in particular, by Jacobi. Can we conclude then that Jacobi had read it? Certainly not, but we can take a small step forward by scrolling the index of that "incriminated" issue of the Crelle Journal: it turns out that, curiously, in that same issue, there is also an article by Jacobi! It was about an unrelated geometry problem, but this is not the point: in light of this observation, we are strongly tempted to make the hypothesis that, having gone to the library to check the correct publication of his article, Jacobi had discovered, with great surprise, the long paper by his rival and, reading it, had immediately understood its enormous relevance, so much so as to absorb completely the concept of inversion as if it were his own, proceeding very quickly in the writing of his new article, which was sent to the Journal on November 18, 1828.

This is, of course, just a hypothesis. Coming back to the bare facts, with the entrance onto the scene of Abel and Jacobi, the theory of elliptic functions definitely took a new course. Legendre, as already mentioned, was more than enthusiastic, so much so that he wrote to Jacobi on February 9, 1828: "I am very pleased to see two young mathematicians like you and Abel successfully cultivating this branch of analysis which was for long the favorite subject of my research and which has not received yet the attention it deserved in my country. With this work of yours, you enter the rank of the best mathematicians of our era." The same enthusiasm can also be found in another letter, sent almost a year later, on April 8, 1829: "Both of you proceed so quickly in all your magnificent speculations that it is practically impossible for someone like me to follow you. I am after all an old man, who has passed the age at which Euler died, an age for which one must fight a certain number of infirmities and in which the spirit is no longer able to adapt to new ideas. I warmly congratulate myself on having lived so long that I can witness this magnificent competition between two equally strong young athletes, who direct their efforts to the profit of science, moving its limits further and further."

In truth, Abel was not so strong and athletic: consumed by tuberculosis, he had died at the age of twenty-seven, just two days before the date of this letter.

Further Readings

E.T. Bell, *Men of Mathematics* (Simon and Schuster, New York, 1986)
A. Stubhaug, *Niels Henrick Abel and His Times* (Springer-Verlag, Berlin, 1996)

Boltzmann. The Genius of Disorder

Ernst Mach's imperious voice rang out in the main hall of the University of Vienna: "Professor Boltzmann, I don't think his atoms exist!" He was standing among the wooden benches of the second row, rather agitated, his long grizzled beard shaken by the movement of his head, a strange light in his very black eyes, and his finger pointed towards that big man at the bottom of the room, near the blackboard, who was looking at him in a puzzled manner from behind two very thick nearsighted lenses. It was January 1897, at the annual assembly of the Imperial Academy of Sciences. For the occasion, physicists, mathematicians and philosophers had arrived from the four corners of the imperial kingdom of Austria and Hungary. The hall was fully packed, with not a few having been attracted by the fame of Ludwig Boltzmann, his oratory and the audacity of his theses: he had not disappointed them. He had just concluded his talk, entirely dedicated to explaining how the pressure and temperature of a gas could be interpreted in terms of the incessant motion of atoms, which he believed was the ultimate component of matter.

The blackboard was full of formulae; in the middle, there was also a large drawing aimed at illustrating the kinematic characteristics of an elastic impact between two spheres; around that drawing, there was a whole festival of differential equations, approximations of integrals, algebraic formulae, and relationships between various physical quantities. Despite their chaotic motion—Boltzmann had argued—very precise predictions can be made about the collective effects produced by atoms: by adopting a probabilistic vision of reality, it was indeed possible to reach a deep understanding of thermodynamics, in particular, of mysterious growth entropy and the inexorable increase in the disorder of the universe.

Ignoring the murmur of those present, Boltzmann continued to look at Mach, undecided as to whether or not to respond to that provocation, and whether to do it with the same vehemence. It was not the first time that the two had clashed: their controversy had lasted for years, and it was a dispute that embraced two different ways of understanding the world, its laws, the ultimate nature of matter, and the very

© Springer Nature Switzerland AG 2020
G. Mussardo, *The ABC's of Science*,
https://doi.org/10.1007/978-3-030-55169-8_2

approach to science. The terrain of that fierce clash stretched far back into the past, even as far as the ancient Greeks: atomism.

The Power of Poetry

The most recurrent theme in *De Rerum Natura* was that of atoms, called *rerum primordia*. At the beginning of the poem, Lucretius openly declared that he wanted to explain the nature of things to Memmo. Bodies are not tight and compact, he said; on the contrary, inside, there are large empty spaces. Any given substance is formed of atoms, but, when it disintegrates, the atoms return to their free state and fly away, invisible and indestructible, immutable and indivisible. To explain the course of events, there was no need to involve the Gods, nor, on closer inspection, even humans! If everything was reduced to the motion of atoms, what appeared to be chaotic movement was, in reality, the result of a strict determinism, in which the future was simply constrained by the equations of motion and fixed once and for all by the speeds and positions occupied by all particles at a given instant. This cleared the field of the whims and cruelty of the Gods and opened the doors to a philosophy that was equally indifferent to the desires and aspirations of men, whose ultimate goal was to eliminate pain and the fear of death. The undisputed master of this philosophical doctrine was Epicurus, who had learned the atomic doctrine from the Greek philosophers Democritus and Leucus. Ironically, Lucretius committed suicide after he ingested a love potion that led him to madness.

Like a Karst River

Despite the suggestive power of Lucretius' verses, the atomistic hypothesis of matter had to wait almost two thousand years to find its solid scientific confirmation. This was a victory that was far from easy, the result of several twists and sudden changes, and also marked by bitter academic controversy. As a matter of fact, Democritus' ideas very soon fell into oblivion, opposed by Plato and Aristotle, who criticized the absence of a first principle capable of guiding the perennial flow of natural phenomena. The world could not be chaotic, Aristotle claimed, because the evolution of a living being is such a complex process that it cannot be the result of simple fortuitous combinations of atoms. This position was adopted by medieval Scholasticism and by the fathers of the Church, who were particularly careful to reject the atomistic conception of matter, since they saw in it the terrible germs of atheism. Furthermore, fully taken up by the problem of proving the existence of God, medieval philosophy abandoned its interest in nature and scientific investigation.

The only exception to that general obscurantist climate was the practice of alchemy by esoteric philosophers, driven by their curiosity to understand how the various substances could be transmuted, especially the base metals into gold. As is known, Newton was the first great modern scientist, but also the last of the "magicians": he was mystical, secret and contradictory. If, on one hand, he came to the formulation of the laws of dynamics, capable of predicting the motion of any object in the cosmos, on the other hand, he spent most of his life dabbling in alchemical practices and reading the Bible. Despite Newton's mechanics, which imposed order upon the incessant motions of the planets and gave rise to a new vision of the Universe, atomism, meant as a scientific and philosophical vision, was never at the top of the list of worries of the great English scientist.

Alchemy, however, gradually progressed, eventually flowing into chemistry, and, with this turn, atomism once again came into vogue. In carrying out their experiments, the first chemists had begun to observe that the various elements were combined according to precise rules of weight: one mole of carbon and one of oxygen gave rise to a mole of carbon monoxide, while a mole of hydrogen and two of oxygen produced one mole of water. Such precise regularity could be explained by the existence of atoms, considered the last, indivisible part of each substance. Chemists of the late eighteenth and nineteenth centuries were not interested in investigating further, to enquire whether the atoms were point or extended objects, whether they moved freely in space or were subject to reciprocal forces: they simply wanted to "keep the accounts" and, indeed, what they had discovered seemed to work, as evidenced by John Dalton's law of multiple proportions.

If this was the point of view of chemists, at least until the mid-nineteenth century, in the field of physics, there had been two great intuitions that had opened up very interesting perspectives on the composition of matter, at least as regarded gases and their state laws. The first was credited to Daniel Bernoulli: his famous 1738 treatise 'Hydrodynamica'—a term that he coined and that has been universally adopted since then to indicate that branch of physics that studies the motion of fluids—had a chapter in which he discussed the kinetic theory of gases. Bernoulli had managed to demonstrate a very interesting thing, namely, that pressure can be interpreted, on closer inspection, as the macroscopic manifestation of the innumerable microscopic collisions of the molecules on the walls of the container containing the gas. Moreover, in that same chapter, Bernoulli carried out a series of arguments in favour of the thesis that, for the same volume, the increase of the temperature in the gas gave rise to a similar increase in the number of collisions, thus coming to predict that the pressure of a gas had to grow with increasing temperature.

A further contribution to the atomic theory of matter came from a priest of Serbo-Croatian origin, a travelling philosopher for various European academic and ecclesiastical venues: Ruđer Josip Bošković. Born in 1711 in the Dalmatian Republic of Ragusa, he had become a Jesuit, but he was less interested in theology than he was in worldly matters, with a deep curiosity in the fields of astronomy, mathematics, physics, and poetry, not to mention the subtle art of diplomacy. During his cosmopolitan and multifaceted life, he published several scientific memoirs, including his *Theoria Philosophiae Naturalis*, in which he emphasized the forces of attraction that the atoms exert upon each other. In modern language, these are nothing less than the van der Waals forces, i.e., the short-range atomic forces necessary to understand the liquefaction of gases. However, Bošković's theory did not take hold: to contemporaries, it seemed too ad hoc an idea, based on a series of purely speculative assumptions and without any scientific content.

Like a karst river, the theory of atomism therefore sunk once again, until it found a new paladin in the nineteenth century in the figure of Ludwig Boltzmann.

The Restlessness of a Viennese Physicist

Born in the Vienna of Franz Joseph, a world outlined by Robert Musil's admirable pen in his masterpiece *The Man Without Qualities*,—simultaneously the end of one era and the beginning of another—Ludwig Boltzmann represented one of the time's most emblematic and significant figures. He can be regarded as the ideal link between classical and quantum mechanics, i.e., between the great scientific tradition of the nineteenth century and the extraordinary developments of the twentieth century. He was responsible for the interpretation of thermodynamics based on statistical mechanics and for the audacious atomic hypothesis of matter, a hypothesis that he defended strenuously throughout his life, but whose definitive success he never witnessed.

Boltzmann was both a passionate scientist and a talented teacher. He had a surprising memory and often lectured without consulting any notes. His explanations were crystalline, witty, full of stimulating anecdotes, and enlivened by unusual expressions, such as "gigantically small" or "momentarily definitive." Lise Meitner, one of his last students, was greatly influenced by such an extraordinary teacher. "His lessons were among the most beautiful and stimulating that I had ever listened to… he was always enthusiastic about what he taught, so much so that we left the classroom with the feeling that a new and wonderful world had opened before our eyes," she remembered years later.

In his philosophy lessons (yes, he also taught a philosophy course!), Boltzmann addressed topics that were, at that time, considered exotic, almost bizarre, such as multidimensional and curved spaces. Thus, he immediately attracted the attention of hundreds of people willing to follow his lectures, as well as that of the Viennese newspapers: the news reached the emperor's ears and materialized in an invitation to the Court for a dinner at the Hofburg Palace. Boltzmann was also an excellent pianist: as a boy, he had taken piano lessons from the great composer Anton Bruckner, and he had an unbridled passion for Beethoven. He was brilliant in conversations, and Musil's description of Ulrich in *The Man Without Qualities* perfectly suited him "He has ingenuity, will, open-mindedness, courage, perseverance, momentum and prudence… I don't want to go further, let's say he has all these qualities." And yet, behind that appearance of a brilliant intellectual at the end of the century, his typically Viennese airiness hid a dark side, an uneasiness, a restlessness that went beyond everything, a deep malaise that gripped his throat and created mountains of ice.

For those who are always ready to find hidden meaning in the world, to see the symbolic nature of life's events, there is perhaps no better example than the birthdate of this great Central European physicist: February 20, 1844, the night of Shrove Tuesday, the day between the Carnival spree and the beginning of Lent. Indeed, throughout his life, Boltzmann always oscillated between these two extremes, quickly passing from moments of great happiness to periods of deep depression. A modern psychologist would consider it a typical example of a bipolar personality. From a clear and diurnal world, there was, in fact, a door in his heart that led to another world, stormy, naked and destructive. Like his atoms, he was always in motion, almost moved by internal agitation, difficult to contain. Once again, reading Musil's words:

> The path of history […] is not that of a billiard ball which, once it has started, follows a certain trajectory, but rather resembles the path of a cloud, that of someone who goes lounging in the streets, and here he is led astray by a shadow, there from a group of people or from a strange cut of facades, and finally reaches a place that he did not know and where he did not want to go. The course of history is a continuous heeling.

In fact, after his doctorate in Physics, obtained in Vienna in 1866 under the supervision of Josef Stefan, Boltzmann remained at that same university as an assistant until 1869, the year in which he obtained the chair of theoretical physics at the University of Graz. In 1871, he was offered the same chair in Berlin, but he felt that it was not the most suitable environment for him, opting in 1873 for the chair of Mathematics in Vienna. But after another three years, he was again in Graz on a chair of Experimental Physics. And it was there that he met Henriette von Aigentler, the first female student of that university, who soon became his wife. They had five children, two boys and three girls. In 1890, he returned to Vienna, and in 1893, upon Stefan's death, he took over the chair of theoretical physics. In 1900, he decided to go to the University of Leipzig, ultimately returning to Vienna in 1902: the Emperor Franz Joseph gave him back his chair under the solemn promise that he would no longer accept other positions outside of the borders of the empire, and so it was.

If, in moments of euphoria, Boltzmann could be light, brilliant, and animated by a wave of reveries and joie de vivre, in dark moments, his soul would split itself: he began to hear voices, rumbles, laughter; the metaphors that he loved to use suddenly became things, the things words, the words boulders. The intolerable tension of his life had previously caused him to pass through the doors of psychiatric institutions. And it was precisely in one of his dark periods that he tragically decided to end his life, tormented as he also was by asthma, the aggravation of blindness, and the frustration that came from the fierce aversion to his ideas that he had experienced in many scientific circles. It happened on September 5, 1906, in a hotel in the small village of Duino near Trieste, a seaside village then on the borders of the empire: it was a tragic end, in spite of the enchanting gleam of the bay below, the beauty of which was celebrated in the verses of Rilke. Ironically, the work that would soon convince everyone of the validity of the atomistic hypothesis had been written just the year before by the most famous employee of a Patent Office that history remembers, Albert Einstein.

A Science with Many Fathers but Few Principles

The tragedy and magnitude of Ludwig Boltzmann's scientific work cannot be understood without taking into account the important scientific developments that took place in the nineteenth century, one of the richest periods in the history of science. If, on one side, the high degree of improvement achieved in the fields of mechanics and astronomy had allowed for the construction of a unitary interpretation of Nature, on the other side, there was a multitude of phenomena—such as heat, light, electricity and magnetism—that seemed to elude that theoretical synthesis. This was the reason why, throughout the nineteenth century, investigations of these phenomena and their relationship with the more familiar mechanical processes assumed a fundamental importance.

A decisive impulse came from the Industrial Revolution that began in those same years, especially in England. Right there, the concept of energy found its full valorization, assuming, together with matter, the role of a fundamental category to which all natural phenomena can be traced: it was the birth certificate of a new science, thermodynamics. With the additional help of some observations previously made by the German doctor Julius Robert von Mayer and, above all, by the French engineer Nicolas Léonard Sadi Carnot, it was, in fact, the English scientists William Thomson (the future Lord Kelvin) and James Prescott Joule who laid the foundations for this new scientific building. Thermodynamics is therefore a science with many fathers but few principles.

The first principle of thermodynamics states that energy, meant as the sum of mechanical work and heat, is conserved in every physical process, regardless of the nature of the process. For the materialists of the time, this was the guarantee that nature is eternal, and it gave voice, once again, to Lavoisier's famous saying: "Nothing is created or destroyed, everything is transformed." But the study of the steam engine, which had contributed decisively to the discovery of the first principle, also revealed a pathological tendency of the way in which things operate in the world, that is, that the mechanical part of energy has a bad habit of dissipating constantly, becoming less and less usable. This observation led to the discovery of a profound and general law, immediately promoted to the rank of the second principle of thermodynamics, which indicates the direction of the flow of time, the irreversible course of natural processes. To measure this irreversibility, a new abstract entity was introduced, entropy, a quantity that always increases (or, at most, remains constant) in each thermodynamic transformation. With the formulation of the second principle, once and for all, the chimera of being able to create a mechanical device that carried out perpetual motion, producing energy at zero cost, was ended. But the same principle also led to the disconcerting conclusion that the Universe—as a big, isolated system—was inevitably destined to undergo a thermal death, or to stop any transformation that was not a pure exchange of heat between its particles.

The Mathematical Law of Chance: Probability

The nature of the physical world is, however, subtle, and must always be investigated in great detail. To better understand how things were and to contrast what seemed an absolute irreversibility of physical processes, we had to appeal again to mechanics and to the ancient theory of atomism, this time, though, reformulated in a probabilistic key! The serious difficulties that seemed to contradict the traditional mechanical conception of phenomena in a totally irreconcilable way with the new science of thermodynamics were, in fact, resolved by a new research path, statistical physics and, in particular, by the kinetic theory of gases, which had its top proponents in Ludwig Boltzmann and James Clerk Maxwell. Continuing the research started by Maxwell, Boltzmann, in fact, showed that, whatever the initial distribution of the velocities of the various molecules of a gas, these velocities—due to the effect of collisions— tend to be distributed according to a universal probabilistic law. During his studies, Boltzmann also introduced a mathematical quantity, a function of the velocities of the particles, and was able to demonstrate that, during the entire dynamic evolution of a system, it can never decrease, in a similar way to what happens with entropy. The concept of entropy itself therefore took on a completely different meaning and the natural tendency of this quantity to grow simply meant that there is a natural tendency for physical systems to move from a state with less probability to one with more probability: the mysterious entropy, in Boltzmann's formulation, became the measure of disorder that is present in the motion of molecules. The precise link between the entropy of a body and the probability of its macroscopic state was expressed by the famous formula $S = k \log W$—which is engraved, in imperishable memory, on his grave in the Vienna cemetery—where S is l'entropy, W is the probability in question and k a constant (called, in his honour, Boltzmann's constant).

The increase in entropy of an irreversible phenomenon, which seemed to elude any traditional mechanical theory, found a natural interpretation in the new statistical formulation by Boltzmann. The introduction of these new categories of thought— order and disorder—changed the cards forever on the table of the physical scenario: in light of statistical mechanics, it was now possible to admit a reciprocal convertibility between an ordered (mathematically less probable) and a disordered (mathematically more probable) state. In this way, the probabilistic character of the second principle of thermodynamics was finally understood.

Statistical mechanics also profoundly changed the traditional conception of the physical knowledge of nature: the mechanical behavior of the molecules, which moved by the billions into the smallest portion of sensitive matter, could no longer be treated individually, but only statistically.

Different Visions of the World

There is thus a deep difference between microscopic and macroscopic scales: if a macroscopic body seems to be at rest, microscopically, there is instead a movement of billions of atoms invisible to human eyes. This gave rise to a sort of dichotomy between sensitive knowledge and conceptual knowledge, a dichotomy that some, in the name of a strictly empirical or phenomenological conception, tended to reject violently. Ernst Mach was one of the most energetic proponents of the purely phenomenological approach to physics: he firmly rejected atomism and ostracized Boltzmann's work in every possible way and on every possible occasion. However, according to Boltzmann, science must not be limited to a simple transcription of data: there is no equation that can *exactly* describe any physical event, and therefore it is always necessary to idealize and go beyond experience. The fact that this is inevitable derives from the very process of our thinking, which consists in adding something to experience and formulating a mental image. Phenomenology should therefore not claim that experience cannot be exceeded, but, on the contrary, should encourage us to do so as much as possible.

The subsequent development of science has revealed all of the flaws that are inherent in the conviction of founding physics on a purely phenomenological basis. The discovery of the discontinuous character of the energy and mass of the electron, the development of ever more refined models of the atom, and the realization in the laboratory of fascinating phenomena such as the condensation of atoms at temperatures close to absolute zero did nothing less than confirm the validity of many of the ideas for which Boltzmann had fought bravely, and in extreme solitude, against the official science of his time.

Further Readings

S.G. Brush, *The kind of motion we call heat: A History of the Kinetic Theory of Gases in the 19th Century* (North-Holland, Amsterdam, 1976)

C. Cercignani, *Ludwig Boltzmann: The Man Who Trusted Atoms* (Oxford University Press, Oxford, 2006)

Chandra. The Journey of a Star

On the morning of August 1, 1930, at the P & O Bombay pier, Peninsular and Oriental Lines, there was the usual confusion of all departures, with the porters loading baggage and large crates, port officials working to make the latest arrangements and sailors who went up and down from the piers between ropes and shrouds. Foreign travellers could be recognized by the admiration with which they looked at the Gateway of India, considered the quintessence of the colony, a monument built by the British in the twenties to symbolize their entry into the Raj, only to bear witness, later on, to their melancholy exit. Just a few hours earlier, the rain brought by the monsoon had swept the docks with furious gusts, but then the heat had returned, a suffocating heat; due to the humidity, clothes were wet and glued to the body, until they had become almost transparent.

Near the pier of the Pilsna steamer of Lloyd Triestino, whose route was from Bombay to Venice, a small crowd of people had gathered, composed of relatives and friends there to see off a boy who was embarking on a journey to Europe. Minute, very dark - in fact, almost black - with a large nose, thick lips and shiny black hair that framed his forehead, that young Indian had the dignified demeanour of a Brahmin and a seriousness that was quite unusual for his age. An air of sadness veiled his gaze; he was missing the most important hug, that of his mother, forced to stay at home because she was seriously ill. It was she, most of all, who had encouraged him to embark on that journey to England and to the temple of science called Cambridge. As he was leaving India, he feared that he would never see her again.

The ship finally left the dock. The silhouette of the large colonial hotel Taj Mahal and all of Bombay soon became a dot on the horizon. That journey across the waters of the Arabian Sea, the Suez Canal and the Mediterranean changed the course of astrophysics. That journey also changed the life of that boy forever: his name was Subrahmanyan Chandrasehkar, universally known as Chandra.

© Springer Nature Switzerland AG 2020
G. Mussardo, *The ABC's of Science*,
https://doi.org/10.1007/978-3-030-55169-8_3

The Legend of Ramanujan

Chandra was born on October 10, 1910, and, as the first male child, he was named after his grandfather, Chandrasekhar, which, in Sanskrit, means "moon." He was born in Lahore, a city of Muslim culture in the north of the Indian continent, known for its mosques and sumptuous monuments to the Mughal era: a very distant place, both geographically and culturally, from the Tamil region of southeast India from which the family originated. They were in Lahore because their father, Ayyar, an accountant for the British colonial administration, was often forced to travel for work and had been transferred there as an employee of the Northwest Indian railway company. In colonial times, joining the British public administration was a very popular choice for Indians, both for the interesting career prospects and for the guarantee of a certain comfort. Chandrasekhar's father was one of those who followed this path, working hard throughout his life to ensure his family's well-being. At home, he dressed in the style of South India, with a comfortable raw cotton shirt, but at work, he adopted western clothing, with shirt, tie and trousers, although he did not give up wearing the typical traditional turban to emphasize his origins. He was very much a man of his time, authoritarian and ungenerous with affectionate gestures: for his daughters, he arranged their weddings, for his sons, he tried to arrange their professional lives. His great desire, in particular, was that his first son should follow his path and find employment in the colonial administration. However, he would face disappointment in this regard.

Chandra was not interested in getting a job in the great colonial machine of the British Empire; he found far greater enjoyment in mathematics. In Madras, where the family had moved in 1923, he loved spending many hours in his home library, a bright room surrounded by mango trees and coconut palms. Here, he had discovered many advanced mathematical texts, as well as volumes full of formulae and diagrams handwritten by his paternal grandfather. In that library, new fascinating horizons had opened up for him, and he had immersed himself in the study of geometry and analysis, so much so that, at fifteen, he had mastered authoritative integrals, derivatives, and all other advanced means of calculation. He was strongly encouraged in his studies by his mother, Sitalakshmi, a woman with an open mind and a sunny nature: "Chandra, do not listen to your father, do not be intimidated, just do what you like." The father had to surrender to the evidence. In 1925, Chandra enrolled at Presidency College, the best university in the country, where he soon became famous for his skill in scientific disciplines.

His idol was Srinivasa Ramanujan, the first Indian to be elected a Fellow of the Royal Society. Ramanujan's life was almost legendary, and he was the very symbol of innate genius. The amazing results obtained by Ramanujan in number theory—the queen of mathematics—had opened up to him the doors of the Trinity College of Cambridge and had attracted the attention of the great English mathematician Godfrey Hardy. Chandra had heard about the great Ramanujan from his mother and uncle, the famous Indian physicist Raman, and the expressions used by both of them were of proud admiration. Those remarks remained indelibly impressed in Chandra's

memory, whose greatest aspiration was to follow in his footsteps, to become a great
mathematician like Ramanujan. But, on this particular issue, the father was extremely
firm: according to him, you do not get very far with Mathematics; if you really want
to continue in scientific studies, you would be far better off turning to Physics, like
Uncle Raman, who had had an amazing career.

Indian Science

That period, in fact, was among the most favourable ones. Indian science, and, more
generally, all Indian culture, including all artistic and literary expressions, experienced a moment of great splendour in the '20s. While Rabindranath Tagore, with
his verses and stories of great lyricism, received international acclamation, men of
science like Satyendra Nath Bose, Meghanad Saha and Raman were destined to leave
an indelible footprint on the history of science. Their recognition by the British, and,
more generally, by the Western world, was a source of inspiration and pride for
millions of Indians: a form of redemption from the British domination, a way to
proudly affirm their skills.

Satyendra Bose, for example, was the first to formulate the hypothesis that particles or atoms with a whole spin, such as the Rubidium, Potassium or Helium atoms,
satisfy a particular statistic. This form of quantum statistics, applied to photons—the
particles from which light is made—immediately made sense of the spectrum of the
black body that had so distressed Max Planck. Nowadays, the statistics of bosons,
particles so named in honour of Bose, are the ones that explain the amazing quantum
effects, such as superfluidity or superconductivity, that occur in many materials at
very low temperatures. Megh Nad Saha, on the other hand, developed an equation,
soon to become famous, capable of predicting the ionization of a gas, and thus
providing a correct interpretation of the stellar spectra. Raman, Chandrasekhar's
uncle, had, in 1928, made the important discovery of the effect that bears his name:
the Raman Effect, which allows for identification of the structure of a molecule
starting from the way in which it interacts with light, opening new paths in the field
of spectroscopy and, in 1930, opening the path to the Nobel Prize in Physics for
its discoverer. This intellectual impulse was also accompanied by all of the political
fervour that was shaking India in those years, with the aim of achieving independence
from the British Empire: Mahatma Gandhi, Jawahartal Nehru, Sardar Vallabhal Patel,
and other political leaders had quickly become equally famous names, both on the
Indian continent and abroad.

The Meeting with Quantum Physicists

Once he chose to study Physics, Chandra studied one of the classics of science in
depth: the treatise *Atomic Structure and Spectral Lines* by Arnold Sommerfeld, a book

in which the principles of spectroscopy and quantum mechanics were first revealed, at least in the version known at the beginning of the '20s, i.e., before the scientific revolution and before the final formulation of quantum mechanics was provided, some years later, by scientists such as Heisenberg, Pauli, Bohr, Schrodinger and Born. Chandra had also had the chance to meet both Sommerfeld and Heisenberg personally, since both visited India in those years. In the afternoons they spent together, Sommerfeld explained to the young Chandra, in great detail, the new concepts of quantum physics that overturned all of the ideas hitherto proposed for understanding the atomic world.

It seemed, for example, that the quantum nature of electrons in a metal unexpectedly produced an outward pressure that could hold the metal together and give it the properties of hardness that characterize it. Impatient as he was to learn all of these new developments, Chandra retrieved the articles by Dirac and Fermi recommended by Sommerfeld and, in a very short time, he wrote his own first article, which was published in 1928 by the Indian Journal of Physics. He decided to make it a more substantial publication, to be published in the prestigious *Proceeding* magazine of the Royal Society, and so he contacted Ralph Fowler for the first time. However, his most stimulating encounter was with Werner Heisenberg, who came to India to give a lecture at Presidency College. At the age of twenty-eight, Heisenberg was already a celebrity for his fundamental discoveries in the field of quantum physics, such as his famous uncertainty principle or the formulation of quantum mechanics in terms of the algebra of matrices: he was considered by many a possible candidate for the Nobel Prize. Although very young, as the best student at the College, Chandra was given the prestigious task of delivering a speech on Heisenberg's work and presenting the guest. "In just a day," Chandra later wrote to his father, "simply talking to him, I managed to learn a world of things on many physical issues. At night, we drove along the Madras marina and he told me about America."

In the summer of the first year of his studies, Chandra was invited by his uncle to spend those months in his laboratory in Calcutta, with the possibility of having his first real scientific research experience. But, as in the case of the famous physicist Pauli, known for his idiosyncrasy in regard to the laboratory apparatus, his experiments were a disaster, the instruments were mysteriously broken at each measuring attempt and nothing seemed to work. It became immediately clear that his true vocation was theoretical physics. He returned to Madras, bringing with him a stimulating book: *The Internal Constitution of the Stars* by Arthur Stanley Eddington. In this treatise, Eddington led the reader to the frontiers of astrophysics at the time, with the brilliant style of a great writer. "It is reasonable to think," wrote Eddington, "that in the not too distant future we will be able to understand something as simple as a star." Chandra was so fascinated by this challenge that, having won a scholarship to study in England, he had no doubt where to go: Cambridge, the temple of science, where Fowler, Eddington, and all of the other great names in astrophysics were. His dream was to understand the stars and become as famous as them.

The Fate of a Star

Although they seem eternal to us, stars are nevertheless mortal. Like everything else in the world, they are born, live, and then die. Depending on the mass, the birth of a star can take millions or hundreds of millions of years, apparently a long time, but still a blink of an eye on the scale of the universe. The stellar incubators are the icy regions of the stellar nebulae—the Orion Nebula, the spectacular Eagle Nebula or the impressive Great Magellanic Cloud—made up of the simplest and most abundant gas in the universe: hydrogen. Accidentally colliding, the gaseous particles of these nebulae begin to accumulate around an initial core and, when the concentration of atoms becomes so dense that the temperature reaches millions of degrees, nuclear reactions begin. The star can thus give way to life. For how long it will then live and how it will die also depends solely on its mass.

The stages are always the same: initially, the star burns the hydrogen, transforming it into helium, and when there is no longer hydrogen, it begins to burn helium, producing carbon. A star of mass like that of the Sun never exceeds the carbon stage, because its mass is not sufficient to raise the core temperature above those one hundred million degrees necessary for carbon to turn into heavier nuclei.

The vast majority of the Milky Way is made up of stars that are colder, smaller and less luminous than our sun and will live billions of years, becoming, in the end, nothing more than white dwarfs, compact cosmic objects that shine with the cold light of diamonds. But for stars of mass larger than that of the Sun, about 1.4 times bigger, their destiny is very different. These stars, once all of their nuclear energy is exhausted, become so dense that gravity can win over every other force, thus bringing the star to a complete collapse. This collapse can result in the formation of a neutron star, but, due to very high masses, it can also give rise to a black hole, i.e., a cavity of space-time in which what remains of a dead star rests. For decades, scientists have rejected this idea by considering it a pure eccentricity, an eccentricity that, in fact, could not exist: nature would have forbidden such absurdity—exclaimed Sir Arthur Eddington—since an ugly solution appeared for the most beautiful theory ever created, that of the general relativity of Albert Einstein. But today, we know that the universe is full of these cosmic monsters, and one of the most gigantic ones is located right in the center of our galaxy. In order for a star to die like a white dwarf, its mass must not exceed a limit value, because, above this, there is only the gravitational abyss: the existence of this value (1.4 times the mass of the Sun), known as the limit mass of Chandrasehkar, was the extraordinary discovery achieved in the cabins of the Pilsna as it sailed towards Europe.

A Surprising Conclusion

After it left Bombay, the Pilsna stopped in Porto Said, and then in Aden. Along the route, herds of dolphins followed the ship, having fun playing by splashing in the water and jumping on the waves. The great Arabian Sea sent wonderful reflections, the same ones that had inspired his uncle Raman towards the great discovery that had earned him the Nobel Prize. Chandra also dreamed of making a sensational discovery, of becoming famous, of returning to India as a great scientist. A crowd of thoughts and emotions filled his head and, to distract himself, he thought there was nothing better than to continue his study of physics. He had with him the article that he had written in Madras, in which he had developed Fowler's theory of white dwarfs. Fowler, using the new laws of statistical mechanics of Fermi and Dirac, had managed to demystify a long-standing paradox about the extraordinary density of white dwarfs, and Chandra now wanted to get a better description of what was happening in these stars. One of the first conclusions he came to was that the density at the center of the star was six times greater than their average density. Suddenly, a fundamental question occurred to him: if the density at the center is so high, is it possible that we should also consider relativistic effects? The point, according to the Fermi-Dirac statistics and the Pauli principle, is that two electrons can never occupy the same state. Thus, taking a certain number of electrons, they tend to fill all available levels, and those with higher momentum have much higher energy than their resting mass. In this case, it is necessary to use the special relativity laws of Einstein for their correct description. Excited by this discovery, Chandra immediately went to work and, to his great surprise, discovered something absolutely unexpected, a stunning and surprising result.

If his calculations were correct, there was indeed a limit on the mass of a star that allowed it to evolve into a white dwarf. If the mass of the star was too large, it could not become a white dwarf, as Fowler's theory would have predicted, in this case, a negative value for its radius! The limit value of mass, on the other hand, was expressed in terms of fundamental constants of physics and the molecular weight of the star. Chandra was fascinated by this result, having been able to find a limit value of the mass of a star in terms of fundamental constants! Fascinated, but confused: what would have happened if the star had had a mass larger than this limit? Would the star disappear into nothingness, engulfed by its own gravity? The idea, as Chandra discovered years later in an episode that would cut him deeply, was not only revolutionary, but also heretical.

Oxbridge

Once in Cambridge, he completely immersed himself in the atmosphere of the colleges, the alleys that led to the riverside, the rainy days, the large bookstores. Together with Oxford, Cambridge was the real heart of the British academic system: Oxbridge was the term used to indicate that world, made of classroom lessons that exuded history, the haughty but deliberately scruffy air of the professors in their Scottish wool tweed, the exclusive clubs, the literary and scientific circles, the snobbery of their members, which constituted an almost impenetrable barrier for everyone else, the bicycles that roamed the streets, the students in togas, the stained glass windows of the enormous libraries. Travelling along King's Parade or Regent Street, Chandra thought of them as the same roads that the legendary Ramanujan had walked years before, and felt he had a great responsibility on his shoulders.

He spent the next four years completely immersed in work for his doctorate. The first year, he attended lessons on quantum mechanics taught by Dirac, who was famous for following the text of his renowned book word by word, but it was worth it in any case. Dirac became his official supervisor, and Chandra thus had the opportunity to better deepen his knowledge: "He was extremely friendly to me, even though it was clear that he was not interested in what I was doing at all. He invited me to have tea in his room at St. John's, and some Sundays we went out of Cambridge together for long walks."

Chandra was the best student in the course of astrophysics and, for this, he was rewarded: he received the congratulations of Arthur Eddington, who invited him to meet up on May 23, 1931. But two days before that date, he received tragic news: "Mom passed away. A big hug." He was devastated. "I am alone in my pain, closed in my room to tears," he wrote to his father. For two weeks, he remained as if paralyzed, and then, gaining strength, dove once again into his calculations, seeking comfort in the formulae and studies. Work was his panacea for loneliness and sadness. He felt the need to get away, at least for a while, from Cambridge: he thus spent several months in Gottingen, with Max Born's group, which afforded him the exciting opportunity to get to know all of the great theoretical physics of the time and to see Werner Heisenberg again.

Trinity Fellow

He returned to Cambridge in early September, but was in a severe crisis: despite the dozen or more articles that he had written in the field of astrophysics, he felt that he was much more attracted to the problems that were shaking theoretical physics, in which everything was in turmoil. Once a problem was understood, another one immediately appeared. He spoke to Dirac with some embarrassment, letting him know how unhappy he was at Cambridge and how he desired to change his field of research. The advice he received was to visit Niels Bohr in Copenhagen.

There, he found a sparkling atmosphere, very friendly and informal, quite the opposite of that in England. He befriended Léon Rosenfeld and enjoyed Bohr's exquisite hospitality. He worked intensively on a problem that Dirac had suggested to him, that of generalizing Fermi-Dirac's statistics, and, after several weeks, he thought that he had solved it: he wrote an article and submitted it to his mentor for judgment. Unfortunately, the answer was not encouraging: Dirac had found a mistake in the reasoning and, therefore, suggested that he not send the article to any journal for publication. Heavily shaken by this experience, he wrote to his father, demoralized: "Six months of hard work gone up in smoke, my work in quantum mechanics was wrong."

The magic of Copenhagen began to dissolve, and with it his enthusiasm for theoretical physics. "I didn't belong to the Copenhagen science community any more than I did to Cambridge," Chandra admitted years later. It was time to return to work on his thesis before he finished his scholarship in August 1933. He completed the job in a few months, further expanding and developing some of Edward Milne's previous studies on stellar structure. The discussion of the thesis, which took place in Eddington's study, in Fowler's presence, was only a formality. The real problem was what to do next: Chandra had no desire to return to India, as his father had been asking him to do for months, but he was willing to stay—even at his own expense—in some European laboratory or research center. He applied, without great confidence, for a Trinity College scholarship, the "dream place" that everyone wanted: this would allow him to stay in Cambridge for another four years, in the famous college where names of the caliber of Newton, Maxwell, and Lord Rayleigh had loved and worked. Fowler advised him not to count too much on the result: the place was open to candidates of all disciplines and the competition promised to be formidable. Yet, the dream came true, against all expectations. The only other Indian ever elected as a Trinity Fellow had been Ramanujan, some 16 years earlier. Chandra regained his self-confidence, and, in a short time, he also became a member of the Royal Astronomical Society and took part with pleasure in the meetings held in London every second Friday of the month.

That Afternoon at the Royal Astronomical Society

In the autumn of 1934, the problem of white dwarfs was enthusiastically taken up again. He certainly now had a better command of stellar structure and quantum statistics. It was appropriate to refine the calculations made four years earlier during his journey from India: he also received strong encouragement from Eddington himself. The old astronomer often stopped by his office to chat, and even made arrangements to provide him with what was considered a real luxury for the time, a mechanical calculator.

The reality, however, was very different: Eddington was, in fact, convinced that Chandra was going in the wrong direction, as became crystal clear in the meeting that took place on January 11, 1935, at the Royal Astronomical Society. After Chandra's talk, the last speaker of the conference was Eddington, who took the floor with a bit more acrimony and arrogance than usual. He was famous for his sharp tongue, but his speech left even those who knew him and who had seen him in action for years stunned.

Intertwining rather intricate reasoning, in which he mixed the theory of general relativity, of which he was held to be one of only three experts in the world, with quantum mechanics, he articulated himself in a sequence of highly effective rhetorical sentences, accompanying the words with blatant, showy gestures. According to him, there was no possibility in physics that electrons had a speed close to that of light; it was simply not something that was foreseen by the laws of relativity and quantum mechanics. "The being born of this union, I believe, is the son of an illegitimate marriage," he declared, thus making a mess of Dirac's theory for which the English physicist, his colleague at Cambridge, had received the Nobel Prize just two years earlier. He then went on, in the same style and with the same arrogance, using sarcasm as a razor blade, making a distinction between physical and mathematical theories. "Dr. Chandrasekhar's idea that a star must contract indefinitely is a cosmic antic. I think there must be a law of Nature that prevents a star from behaving in such an absurd way," Eddington proclaimed at the conclusion of his speech.

The blow was devastating. Chandra knew that he had been publicly ridiculed by the most respected astronomer of the time. And it wasn't even the only time that this happened. There was a void around Chandra, and in the following years, there was no scientist who was willing publicly to speak a word in his favour. In a letter to his father from 1940, he wrote, "I feel that all astronomers, without exception, think that I am wrong. That I am some kind of Don Quixote trying to cripple Eddington. It is a very discouraging experience. The only alternative is to change my field of research."

Along with white dwarfs, Chandra also left the British Empire behind. He emigrated to the United States in 1936, accepting a position as a professor at The Yerkes Observatory, near Chicago. His calculations were, however, correct. Eddington's foolish and baronial opposition, in some way inexplicable, had the effect of slowing down the progress of astrophysics for decades. When Chandra won the Nobel Prize in 1983, more than fifty years had passed since his discovery on the Pilsna. The only thing that remained of that young, twenty-year-old man, full of hopes and dreams, was his melancholic gaze.

Further Readings

M. Begelmann, M. Rees, *Gravity's Fatal Attraction: Black Holes in the Universe* (Cambridge University Press, Cambridge, 2010)

K.C. Wali, *Chandra: A Biography of S* (The University of Chicago Press, Chicago, Chandrasekhar, 1990)

Dirac. Beauty Is Truth, Truth Beauty

Robert Oppenheimer once said that Wolfgang Pauli was the only person he knew who was equal to his caricature. Several years later, Jeremy Bernstein felt compelled to add that if Pauli was similar to his caricature, Paul Dirac was even more picturesque than his caricature! The impenetrable reserve, the austere lifestyle, an almost impenetrable secrecy, all combined with an almost obsessive paucity of words, have made Dirac an inexhaustible source of anecdotes and jokes.

An example? He won the Nobel Prize for Physics in 1933, and his first reaction was to reject it, fearing the wave of publicity that would fall on his shoulders with such recognition. But he quickly changed his mind when Ernest Rutherford, his colleague at Cambridge, pointed out to him that, if he refused it, the publicity would be much greater!

An Oppressive Father

Paul Adrien Maurice Dirac was born in Bristol on August 8, 1902, the same year in which Sir Arthur Conan Doyle published *The Hound of the Baskervilles*, a famous novel in the Sherlock Holmes series, and Albert Einstein started working at the Berne Patent Office. Dirac's father Charles, originally from Monthey—a country in the canton of French-speaking Switzerland—had long settled in that city with a rich maritime past in the southwest of England, where he got a job as a French teacher in a local college. Years earlier, he had abruptly left his family in Switzerland, having found it impossible to live with his tyrannical father, and he eventually stopped even writing home to give news of himself. At the Bristol library, Charles met Florence Hannah Holten, a beautiful girl of Welsh origin, very sweet both in face and nature: in 1900, the couple's first child was born, Reginald, and two years later, it was Paul's turn, while the third, a daughter, Beatrice, was born in 1906. As is well known, family stories and dramas tend to re-emerge and repeat themselves: although he had personally experienced what it meant to live with a despotic and prevaricating father,

© Springer Nature Switzerland AG 2020
G. Mussardo, *The ABC's of Science*,
https://doi.org/10.1007/978-3-030-55169-8_4

when it was his turn, Charles became an unpleasant and authoritarian patriarch as
well. He did not hesitate to impose strict rules on the family, to sever all kinds of
social relationships and to show no tenderness towards his children. The atmosphere
in the house was more like that of a barracks or a prison than of a family nest. As Paul
Dirac later recalled, the father demanded that only French be spoken in the house,
punishing those who made grammatical or pronunciation errors: "Realizing that I was
unable to express myself well in French, I decided that it was better to remain silent.
In this way, I became a very taciturn person." The dinners often saw the father and
Paul, alone, in the dining room, engaged in a dialogue conducted mostly in French
monosyllables, especially from Paul, while the other two children and the mother
sat separately in the kitchen, speaking in English (as a child, he was convinced that
men and women spoke different languages). The situation could not help but take a
certain rapid turn, and the young Paul soon became extremely introverted, detached,
almost ascetic, solitary and of very few words: in short, strange, "the strangest man
in the world," in the words of his biographer Graham Farmelo.

The gloomy family atmosphere unfortunately also resulted in drama. In addition
to the daily harassment suffered during his adolescence, there was also the matter
of his education. In that era, when it was time to go to university, male children
were forced by their fathers to study Engineering: Reginald, interested instead in
medicine, took a path of disappointment that led him first to a deep depression and
then to suicide. Paul, contrastingly, passively accepted the choice. However, he found
refuge in a solitary and silent world whose center of equilibrium was mathematics,
with its clear rules, the harmony of demonstrations, the beauty of stringent logic. The
fear that he felt as a boy towards his father turned, as an adult, first into contempt
and later into a glacial detachment: when he went to Stockholm to receive the Nobel
prize, he was accompanied only by his mother, making sure not to invite his father
to the ceremony and, when Dirac père died, Paul wrote to his wife: "I finally feel
free."

The heavy emotional burden of French-based dinners also caused Paul not to be
particularly fond of that language. Among the various stories that have flourished
over the years, one of the best has, as its protagonist, a French physicist who barely
spoke English: visiting Dirac, he was having great difficulty finding the right English
words to correctly express his ideas. While he was babbling, Beatrice, Dirac's sister,
entered the room and addressed her brother in French, which he answered impeccably
in the same. When the astonished guest demanded, "But why didn't you tell me you
speak French?", Dirac's simple answer was, "Because you didn't ask me!"

The Laws of Chances

In 1921, Paul Dirac graduated with a degree in Engineering, but, due to the economic
depression of those years, finding work proved to be a futile undertaking for him.
However, he then obtained a scholarship in mathematics, which allowed him to spend

two more years at the University of Bristol, where he distinguished himself as one of the most brilliant students in his course. The turning point came in 1923, when he moved to Cambridge, a place of great scientific tradition.

He was welcomed at St. John's College, one of Cambridge's wealthiest and oldest universities, which had a large park that was crossed by the River Cam. It had been founded in the sixteenth century as the Trinity College or the King's College: the architecture was solemn, dominated by the chapel, with its majestic Gothic windows, ancient wooden benches and imposing organ. Inside, there was a large courtyard surrounded by the various wings of the building, with studios for the fellows, reading and tea rooms, the dean's office, and pantry rooms. There was also a beautiful library, full of leather-bound books, the great manuscripts of the past, endless rows of volumes dedicated to mathematics, geometry, physics, etc. On the walls of the huge refectory hung portraits of the greatest academics associated with the college, the Johnians, all with serious faces and academic robes; some seemed to be looking you right in the eyes: among them were the famous writers William Wordsworth and Samuel Butler, the famous Prime Minister Lord Palmerston, and many others.

For Paul Dirac, it was a totally new atmosphere compared to what he was used to in Bristol. Life inside of the colleges reflected the historical tradition of those places, when only the monks studied, the day outlined by learning and prayer. Despite some changes, the general atmosphere had remained the same: college fellows had to get up early in the morning and wash themselves with cold water; afterwards, they would go to the large common room to have breakfast, and then immediately run off to study. In the evening, they would all gather together again for dinner, with the dean and the most eminent personalities sitting at the high table and marking the time dedicated to meals. Everyone was expected to have a conversation with his neighbours at the table. But having a conversation with Dirac was practically an impossible undertaking: in order not to be forced to exchange even a word, he would not even ask for the salt. Once a diner in front of him asked him, "It's raining, isn't it?" Dirac got up from the table, went to the window and looked outside, then returned to his place and said laconically, "No." That was all.

He loved the splendid isolation of his room, as well as the silence of the library and the vast lawn that stretched along the banks of the river. When he arrived in Cambridge, his scientific interests were all directed towards the most mysterious force present in nature, the force of gravity, and to the suggestive interpretation that Albert Einstein had given it a few years earlier. He had come to Cambridge for the express purpose of deepening his study, but Ebenezer Cunningham, the leading local expert in the theory of general relativity, made it clear that he was not taking on any other students. It was then that chance stepped in and decided Dirac's academic career: he was assigned, as his supervisor, ex officio, Ralph Fowler, and this was to his great fortune.

Indeed, Fowler was the ideal link between the experimental and theoretical physics of the time, certainly the greatest exponent in Cambridge of the nascent atomic theory. Thanks to him, the doors of that new and fascinating world, full of surprises and challenges, opened up for Dirac: "Fowler opened my eyes to the atom, to the studies of Rutherford, Bohr and Sommerfeld. Before then, I had never heard of the Bohr atom, and it was a great surprise to me to discover that the laws of electrodynamics did not apply to the atom. Atoms had always been considered hypothetical entities, and now there were many physicists who dealt with their structure. I concentrated completely on scientific work and continued to dedicate myself day after day, except on Sundays, on which I rested: if the weather turned beautiful, I loved taking long walks alone in the countryside."

In just two years, he managed to master that strange quantum world, and in the following ten years, he changed our vision of the everyday world forever. As is well known, physical reality rests on two large pillars: the theory of relativity and quantum mechanics. Dirac was one of the founders of the second; in addition, in trying to make it compatible with the first, he came to one of the greatest discoveries of all time, the existence of antimatter. Therefore, it is not surprising that, together with Newton, Maxwell and Einstein, Dirac is rightly considered today among the most important physicists of all time.

It all began in the summer of 1925, when the German Werner Heisenberg, at the time twenty-four years old but already the undisputed rising star of the world of theoretical physics, was invited to Cambridge to hold a seminar within the activities of the Kapitsa Club, a scientific forum particularly interested in scientific advances. The topic of the seminar was quite technical: atomic spectroscopy in light of the formulation that had been given by Bohr and Sommerfeld's theory. It was only at the end of his speech that Heisenberg mentioned, almost *en passant*, the pioneering results that he had obtained earlier in the field of quantum mechanics. In Cambridge, no one was aware of these developments—the related article had not yet been published— and only Fowler managed to get slightly more detailed information during a private conversation with Heisenberg, in which he invited him to send the manuscript as soon as possible. A week later, the article in question was on his table, and Fowler passed it on to Dirac, asking for comment.

What happened next was spectacular: in reading it, Dirac was initially struck by an observation reported in the text, according to which, in the field of atomic physics, it was necessary to contemplate the non-commutativity of the product of two physical quantities. It meant that A × B was not necessarily equal to B × A, an event that seemed to worry Heisenberg himself. This happened in particular if, for A, the position q of a particle was taken and, for B, its mechanical impulse p ... It was a real oddity: "Is it possible that measuring the position of a particle can alter the measure of its impulse, and vice versa?" This eccentricity, however, seemed to move in the right direction; the experimental observations said so... The problem drew the obsessive attention of Dirac. He thought about it and pondered it over and over, until, during one of his usual Sunday walks, he had an illumination, and everything suddenly became extraordinarily clear and simple: the non-commutativity was not only not a problem, it was indeed the essence of the new quantum mechanics!

Years later, he commented that scientists who propose a new idea tend, in general, to have a highly emotional attitude towards their discovery, dictated by the fear of making a mistake or discovering that they have made a mistake: "Lorentz did not have the courage to get to the theory of relativity, Heisenberg, on the other hand, was afraid of the non-commutativity of the products ... Very often, those who have glimpsed these new possibilities are not the best persons to develop them."

His flash of genius was to realize that, in quantum mechanics, the commutator of two physical quantities was a mathematical quantity that was curiously similar to the famous Poisson brackets used in classical mechanics. The latter provide a refined version of the laws of motion discovered by Newton and have the advantage of highlighting a very sophisticated mathematical and geometric structure of the familiar Newtonian mechanics (the one that is usually applied to predict the motion of macroscopic bodies). That day, however, during his Sunday walk, he could not remember all of the aspects of this formulation in detail, and therefore had to wait anxiously for Monday morning—at the opening of the library—so that he could consult the books by Hamilton and Hertz on classical mechanics. His intuition turned out to be completely correct. He quickly wrote an article with the suggestive title 'The fundamental equations of quantum mechanics,' which was published in the *Proceedings of the Royal Society*, the leading journal of British science: he showed that the new rules relating to the non-commutativity of quantities in Physics naturally incorporated the two formulations proposed by Heisenberg and Schrödinger for quantum mechanics, and that they are, in fact, equivalent. The article was read with extreme attention at Göttingen, by Max Born, Pascual Jordan and Werner Heisenberg himself. Also, the three great German scientists had, in fact, understood the importance of non-commutative quantities, associating them with physical observables expressed not by simple real numbers, but by tables of complex numbers, the so-called matrices. Put simply, there was a fierce scientific competition going on between the lonely British physicist Paul Dirac and the German physicist in Gottingen. Fortunately, it was a race marked by fair play: Born expressed all of his admiration for the work done by the young stranger, and Heisenberg felt that he had to congratulate Dirac directly, pointing out, however, "I hope you won't be disappointed if I tell you that an important part of your work has already been discovered here in Göttingen some time ago and recently published by Born and Jordan. This does not take away any merit from your work, the formulation of which is somewhat superior to that achieved here." Dirac was not depressed, on the contrary, he understood that the concomitance of these discoveries, made almost simultaneously there, in England, and on the Continent, showed that he was on the right path towards understanding that strange world of quantum mechanics, no longer regulated by the determinism and certainties of the old, familiar Newtonian mechanics, but by indeterminacy and probability.

When he arrived in Cambridge, it was by chance that he came into contact with quantum mechanics; curiously, his studies, together with those of Bohr, Heisenberg, Schrödinger, Pauli, and many others, showed unequivocally that probability, or the laws of chance, was the true and only conceptual basis upon which one could build the new atomic physics. Of all of his many scientific discoveries, that of the profound link between the commutators of quantum mechanics and the classic Poisson brackets always remained, for Dirac, one of his favorites.

Anti-matter

Towards the end of 1926, Dirac decided to spend a year abroad, visiting the major European centers where quantum mechanics was born. He first spent several months in Göttingen, where he met the most brilliant minds of the time: in addition to the great names already mentioned, there was also Robert Oppenheimer, with whom Dirac became a great friend, in Dirac's own unique way, of course! From the point of view of temperament and passions, Oppenheimer was, in fact, as far away from Dirac as one can imagine: if the former enjoyed French literature, Oriental philosophy and art, the latter's only interests consisted of physics and travel; reading, he said, distracts from thinking. At the time of their meeting, Oppenheimer was immersed in his first reading of Dante, and Dirac once addressed him with the following comment: "How can you do physics and poetry? In physics, we try to express in a simple way things that nobody knew before. In poetry, however, it's quite the opposite!" Over the years, his interest in literature was always very low: when, later on, the Russian physicist Kapitsa gave him *Crime and Punishment* to read, Dirac's only consideration was somewhat surprising: "Not bad, but in one chapter, the author made a mistake: he said that the sun rose twice on the same day."

After Göttingen, Dirac moved to Copenhagen, where he found himself very comfortable in the friendly and informal atmosphere of the group that surrounded Niels Bohr. The Danish physicist was a very engaging type, and he liked to talk: when he finished a research study, he used to dictate the text for the article to his young collaborators, but this always ended up being a great source of torment, because Bohr often changed his mind, never finding peace. Dirac, once witnessing one of these scenes, commented, "They taught me as a child that you don't start a sentence if you don't know how it ends."

Once back in Cambridge, great recognition awaited him: in 1927, he was elected as a Fellow of St. John's College; in 1930, he was appointed a member of the Royal Society, and immediately afterwards, he was offered the prestigious Lucasian chair originally occupied by Newton. These important acknowledgments were due to a fundamental discovery that he had made in 1928, in which he not only gave reason for the intrinsic rotation properties of the electron (the so-called quantum spin number), he even predicted the existence of anti-matter.

It was a winter evening when, sitting at the desk in the office that he occupied in the right wing of the college, with his long legs stretched out towards the fireplace, and immersed in deep reflections on how to reconcile the principles of quantum mechanics with those of special relativity, he stumbled upon the idea that it was necessary to treat the spatial and temporal coordinates of a system on the same level: in his famous 1905 article, Einstein had, in fact, shown that there is neither absolute space nor absolute time, but that they both equally enter into the fundamental equations of physics, like those of Maxwell, for example.

Literally taking the space-time unification proposed by Einstein, Dirac thus arrived at a quantum equation for the electron, now known universally as Dirac's equation: this implied two solutions, one with positive energy, the other with negative energy.

Each of them was also two-dimensional, perfectly accommodating the two polarization states associated with the electron spin. But what did the negative energy solutions correspond to? This was a real puzzle, solved by Dirac only in 1930, with the hypothesis that the negative energy solutions corresponded to particles of equal mass but of opposite charge to the electron, a type of particle not yet discovered.

However, in 1932, Carl Anderson, an American physicist who studied the traces of electrons present in cosmic rays while they were crossing a very intense magnetic field, realized that half of the traces deviated in exactly the opposite direction to the one expected! It was experimental proof of the existence in nature of the positron, the name by which the antielectron is known today. Dirac's hypothesis, a prediction made by blindly believing in the power of mathematics to describe the world, thus proved to be true.

"No other physicist," wrote Helge Kragh, "has been more obsessed than Dirac with the concept of beauty inherent in the equations of nature." To his colleagues, theoretical physicists like himself, he used to say:

"Let yourself be guided by mathematics, at least at the beginning. Mathematicians usually play a game whose rules they invented, while, for physicists, the rules are already given by nature, but the more one goes on, the more one realizes that the rules that the mathematician finds interesting are the same chosen by Nature! We can say that God is an extremely sophisticated mathematician and that he used extremely advanced mathematics in building the world."

When he visited the University of Moscow in 1956 and was invited to describe the essence of his scientific approach, Dirac took some chalk and wrote on a blackboard: "A physical law must possess mathematical beauty." The epigraph still stands there today.

Further Readings

G, Farmelo, *The Strangest Man in the World: The Hidden Life of Paul Dirac, Quantum Genius* (Faber and Faber, 2009)

G. Gamow, *Thirty Years That Shook Physics* (Dover Publications, New York, 1996)

Erdös. The Wandering Mathematician

Given the characters in question, this story could not have ended otherwise. A story that could be called *The Great Misunderstanding*, and its sequel, *The Great Cold*. The protagonists were two giants of twentieth-century mathematics, but it would have been difficult to find two men so profoundly different, so siderally distant. Their names were Paul Erdös and Atle Selberg.

Central European Atmospheres

Paul Erdös was born in 1913 in Budapest, at the time, a sparkling city full of cafes and theaters, whose women were appreciated for their beauty and the gentlemen for their gallantry. The city was dominated by the imposing castle on the Buda hill and the majestic neo-Gothic-style Parliament building placed alongside the Danube. Both of Erdös's parents were high school mathematics teachers: the family's original surname, Engländer, had been changed to Erdös along the lines adopted by many other Jewish families in Hungary. During the years prior to the First World War, life was relatively comfortable for Hungarian Jews: as the mathematician Vázsonyi recalled, "the Habsburgs were not truly anti-Semitic, which, in Hungary, meant that they did not detest Jews more than necessary." Those were the last years of the magical world of Kakania, the empire of Austria-Hungary (k.u.k., that is, kaiserliche und königliche, imperial and royal) described with great irony by Robert Musil in his novel *The Man Without Qualities*:

> Liberal in its constitution, it was administered clerically. The government was clerical, but everyday life was liberal. All citizens were equal before the law, but not everyone was a citizen. There was a Parliament, which asserted its freedom so forcefully that it was usually kept shut; there was also an Emergency Powers Act that enabled the government to get along without Parliament, but then, when everyone had happily settled for absolutism, the Crown decreed that it was time to go back to parliamentary rule…. And in Kakania, at least, it would only happen that a genius would be regarded as a lout, but never was a mere lout taken, as happens elsewhere, for a genius.

© Springer Nature Switzerland AG 2020
G. Mussardo, *The ABC's of Science*,
https://doi.org/10.1007/978-3-030-55169-8_5

The war, however, and the years immediately after, wiped out that joyful atmosphere and that typically Central European lightness, and things changed radically for the Erdös family: in 1915, the father was taken prisoner by the Russians and sent to Siberia, from where he returned six years later; in 1919, Miklós Horthy, ex-admiral of the Austro-Hungarian Navy, after seizing power with a military coup, quickly established the first European fascist regime, characterized by murder and torture. Thousands of people were killed, especially Jews. Paul Erdös's mother had kept her life, but lost her job, while various people, including Edward Teller, John von Neumann, Leó Szilárd and Eugene Wigner, fled first to Germany and then, with the advent of Hitler, to the United States, where they would all take part in the construction of the first atomic bomb.

The Great Book

As a child, Paul immediately stood out for his outstanding mathematical skills. He would multiply three-digit numbers in his mind, solve curious problems that he invented himself ("How long would it take a train to reach the Sun?"), and ask thousands of questions, altogether showing an unparalleled intellectual liveliness. The problems of everyday life were always taken care of by his very attentive mother: this is why Erdös was later, and throughout his life, unable to tie his shoelaces, spread butter on a piece of bread or even peel an orange, i.e., to engage in all of those little practical and social skills of survival that are part of the baggage of everyday human existence. But Erdös was anything but an ordinary man: his hands were always moving, his walk emphasized his lankiness, he was always ready to ask questions, to solve new problems, to make of mathematics a rare mixture of science and art, as demonstrated in *The Glass Bead Game* by Hermann Hesse.

By the end of his teenage years, he was spending his afternoons meeting up with other Jewish mathematicians—the various Szekeres, Vázsonyi, Klein or Turàn—in a city park on the hills of Budapest, where they talked about such things as politics, but, above all, they loved to immerse themselves in deep discussions of mathematics. Fearing that there were spies everywhere, Erdös began to develop a language of his own, made up of metaphors, double meanings, and puns, which sometimes only those who had mastered physics or mathematics could understand: children, or small things in general, were called "epsilon"; grandchildren, "epsilon squared"; communists were "those with long wavelengths," because of the place occupied by red light on the electromagnetic spectrum; music was simply referred as "noise"; to ask someone his date of birth, he used to say "when did you arrive?"; "Preaching" meant to hold a mathematics conference; "SF," aka the "Supreme Fascist," or the "Number One Up there" stood for God, that being who kept the very refined solution to all mathematical problems to himself, written in the "Great Book," a text that he acknowledged would have made him very happy to be able to look at. "This comes from the Great Book" was the utmost compliment a mathematician could receive from Erdös. His spoken English was legendarily incomprehensible, having learned it from his father,

who, in turn, had learned it by reading English texts during his Siberian imprisonment. At the age of twenty, he became internationally famous with his elementary proof of a very suggestive theorem concerning prime numbers, that is, between each integer n and its double 2n, there is always a prime number p, the theorem expressed by the following inequality:

$$n \leq p \leq 2n$$

For Erdös, prime numbers were his faithful companions throughout his life, and the ability to find elementary proofs for many fundamental mathematical theorems became an art for him.

The Man Who Came in from the Cold

Atle Selberg was born in 1917 in icy Norway, in Langesund, a small town surrounded by woods, famous more for its port than anything else. His father was a professor of mathematics in the local high school, where his mother had been a teacher as well. The last of nine children, from an early age, Atle demonstrated an extremely sharp mind and a rather extraordinary mathematical ability: in his home, there was a well-stocked library of scientific books, and reading an abundance of mathematical texts remained his favorite activity for a long time. He would read in every spare moment, even on the train journey to school. This is how he learned to master the laws of analytical geometry and to solve any algebraic problem. He was strongly impressed by a mysterious identity, which he found by browsing a mathematical text written by Carl Størmer: it involved the famous number π and an infinite series made up of all of the odd numbers

$$\frac{\pi}{4} = 1 - \frac{1}{3} + \frac{1}{5} - \frac{1}{7} + \cdots$$

As he recalled years later, "I understood that I had to read Størmer's entire book to come to terms with such a strange and fascinating formula." A marvelous world had opened up before his eyes and, towards the end of high school, a further incentive to join that world came from his brother Sigmund, already a mathematics student at the University of Oslo: Sigmund brought home an incredible text, a book by the legendary Ramanujan, the great Indian mathematician famous for his mysterious direct line with profound mathematical truths. Atle completely immersed himself in reading this book, in which unimaginable formulas of number theory unfolded their beauty page after page and equations chased one another, bringing to light amazing properties of infinite products, modular functions or transcendental numbers, in a seemingly endless flow. In addition to the myriad formulas, Atle was also deeply impressed by the singular biography of its author, who died of tuberculosis in complete poverty at the age of 33 in his native Indian village, where he had returned after spending several years at Cambridge. As a matter of fact, despite all of the honours and awards that had been bestowed upon him—such as having been elected a Fellow of the Royal Society, or the chair offered to him at the prestigious Trinity College, or the extraordinary collaboration with the illustrious English mathematician G.H. Hardy—Ramanujan had never adapted to the English climate, its food, or the social conventions of the place, and, full of nostalgia for India, he therefore preferred to return to his own country. Reading that book, Selberg decided that he would dedicate his life to the study of number theory.

After high school, he enrolled in the Mathematics Faculty of the University of Oslo, where he became personally acquainted with Carl Størmer, whose book had influenced him so deeply. He was immediately noticed for his skill, and was invited, while he was still a student, to make a presentation at the International Mathematics Conference held in Helsinki in 1938. There, he came into contact with many important mathematicians, including Harald Bohr, brother of the famous Niels Bohr—the father of quantum mechanics—and Arne Beurling, who, at the conference, discussed the Prime Number Theorem, leaving a strong impression on the young Atle.

In April 1940, preceded by a launch of paratroopers, the Wehrmacht invaded Norway and quickly defeated the Norwegian army, giving rise to an occupation that continued until Germany surrendered in 1945. Selberg joined the country's active resistance movement. This did not prevent him from pursuing his research and finishing his thesis, which was focused on one of the most important issues in number theory: the distribution of the zeros of the Riemann function. He was arrested immediately after the discussion of his thesis and held prisoner in Trandum, freed months later only under the condition that he would leave Oslo and return to Gjovik, a Norwegian village to which his family had moved in the meantime.

Birds and Frogs

Freeman Dyson wrote that "some mathematicians are birds, others are frogs. Birds fly high in the air and survey broad vistas of mathematics out to the far horizon. They delight in concepts that unify our thinking and bring together diverse problems from different parts of the landscape. Frogs live in the mud below and see only the flowers that grow nearby. They delight in the details of particular objects, and they solve problems one at a time." We could also say that mathematicians belong, in a first approximation, to one of the following categories: the first includes all of those who believe that solving mathematical problems helps one to better understand mathematics itself; the second, however, is made up of those who believe that a better understanding of mathematics is the best way to solve mathematical problems.

Those who belong to the first category can be called *problem solvers*, while those who belong to the second are *mathematics theorists*. More specifically, it is the difference between Euler, the most prolific of mathematicians, and Gauss, the prince of mathematicians. Of course, in reality, there has never been such a clear boundary between the two categories: in fact, we often see encroachment between the two fields, which obviously mixes the cards and makes the whole thing much more nuanced. According to G.H. Hardy, "a result can be trivial or beautiful, a demonstration can be sloppy, surprising or descending directly from the Big Book. In a good demonstration, there is a very high degree of unpredictability, combined, however, with the inevitability and the economy of thought. The arguments may take a strange and surprising form, the weapons used may be of a childlike simplicity compared to the extent of the results obtained," which shows that the beauty of mathematics is reflected in every fundamental theorem, and vice versa. But, nevertheless, a fundamental difference between the two different approaches to mathematics exists and is perceptible. Einstein once confessed to his collaborator Ernst Straus that the reason that he had chosen to devote himself to physics rather than mathematics was that "mathematics is so full of beautiful and attractive questions that it is easy to waste energy by chasing them without discovering the central ones. In physics, instead, the central problems are very clear and the first duty of a scientist is to go after them, without being seduced by any other problem." And if Paul Erdös can certainly be considered a great champion among problem solvers (as Straus wrote, "he systematically and successfully violated all of Einstein's prescriptions. He succumbed to the seduction of every problem he encountered and many succumbed to him"), there is no doubt that Atle Selberg instead belongs to the other club, that of mathematical theorists.

In 1947, Selberg, in the wake of the results that had made him internationally famous in the field of number theory, left Norway to go to the Institute for Advanced Studies at Princeton, where he remained for the rest of his life, producing works, few in number but fundamental in nature, that profoundly changed the history of mathematics and led him to win the Field Medal in 1950, the highest award for a mathematician. As a result of his reserved nature, or perhaps the discretion typical of Scandinavian culture, Selberg saw mathematics as a research to be carried out in perfect solitude, with the times and methods necessary to penetrate all of its secrets: throughout his long career, he wrote a single article in collaboration and, as he himself said, that was just because it was the mathematician "Chowla who approached me with a question. If he had never come to my office, we would never have written anything together."

Another Roof, Another Proof

As for Erdös, he left Hungary in 1934 for England, thanks to a scholarship from the University of Manchester. From that point on, he became a living legend, not only for his fame due to his ingenious solutions to dozens of problems related to number theory, graph theory, probability, combinatorial laws and infinite other themes, but also for his extravagant way of life: as a perfect wandering Jew, his life was, in fact, all a wander, to the point that he rarely slept in the same bed for more than a week.

In the coming years, he set things up in such a way that he could devote all of the necessary time and energy to mathematics: he never married, he didn't develop any kind of hobby, he never tied himself to any place, he had neither a house nor a fixed position, and all his belongings were kept in a pair of suitcases. He gave most of his money to charity, especially to young, penniless mathematicians, keeping only what was strictly necessary for himself. He used to say, "The French socialists have claimed that private property is theft; I think it's just a nuisance." Always dressed in creased clothing, his face dominated by a heavy pair of black-framed glasses, his appearance frail and sickly, he used to show up at the door of one of the many mathematicians around the world with the ritual phrase, "My mind is open," as a way to say that he was ready to engage in mathematics: once settled in the home, he forced the collaborator to work at a crazy pace, and this went on until he thought it was worth it or until the guest had exhausted all mental energy or physical endurance. At that point, he left, only to land in another house, in another country, or on another continent. "Another roof, another proof," he loved to repeat.

He thus became one of the most prolific mathematicians in history: at the end of his life, he had written 1485 mathematical articles with 509 different mathematicians! No one has ever had a higher number of collaborators than he, and in this regard, somewhere between serious and facetious, the famous "Erdös number" was introduced, a number that, possibly, expresses the level of proximity of a mathematician to Erdös himself: those who wrote a paper together with Erdös have an Erdös number of 1, while those who published a paper with someone who had worked with Erdös have an Erdös number of 2, and so on. Obviously, all of those who have never written any mathematics articles find themselves with an Erdös number equal to infinity! Among his countless collaborators were the Polish mathematicians Stam Ulam and Mark Kac, Ernst Straus and his old friends from Budapest, Alfred Rényi and Paul Turan. And Turàn himself was, indeed, the involuntary cause of the accident that divided the paths of Paul Erdös and Atle Selberg forever.

It Happened at Princeton

On what happened at Princeton in June 1948, we now have the report of an exceptional witness, Ernst Straus, together with the letters written by Atle Selberg to Hermann Weyl, and the correspondence exchanged between Selberg and Erdös: basically, we have all of the tools we need to figure out the sequence of events that took place that summer.

It all began with a formula obtained at the beginning of 1948 by Selberg: this concerns a very important property of the prime numbers p, and, in the following, we will simply call it "the fundamental formula," expressed by the following:

$$\sum_{p<x} \log^2 p + \sum_{pq<x} (\log p)(\log q) = 2x \, \log x + O(x)$$

In June of the same year, Selberg was about to leave Princeton to go to Canada to get his visa renewed in order to prolong his stay in the United States.

It happened that, in the same period, Paul Turàn, a close friend of Erdös's and a great theorist of numbers, was visiting the Institute for Advanced Studies. Strongly intrigued by the fundamental formula, he asked Selberg for some details on the same, as well as whatever notes were available. Selberg, assuming that Turan would be leaving shortly, and thus would no longer be there upon his return from Canada, revealed to him his work and the details of that formula. When he returned to Princeton, ten days later, he learned that Turan, who was very enthusiastic about the mathematical value expressed by the fundamental formula, had, in the meantime, held a lecture on the subject that was attended by Chowla, Straus and Erdös, who had just arrived in Princeton: in particular, during the lecture, Erdös had stated that the fundamental formula could be used to prove another crucial property of prime numbers, namely, that the ratio of two consecutive prime numbers goes to 1:

$$\lim_{n \to \infty} \frac{P_{n+1}}{P_n} = 1$$

Erdös also pointed out that this identity, which, in the following, we will call "*the formula of the ratio*," would lead directly to a completely new proof of the prime number theorem: this theorem, one of the cornerstones of mathematics, was proved for the first time in 1896 by Jacques Hadamard and Charles de la Valleé-Poissin, but only by using all of the artillery and sophisticated instruments of complex analysis available to them. Since the ingenious proof of his fundamental formula given by Selberg was of a far simpler nature, according to Erdös, the path was now opened for a resounding result: the prime number theorem could be proved for the first time in an elementary way! At the first chance, Erdös spoke directly to Selberg about this amazing possibility, but the latter's reaction was decidedly cold, or, more precisely, irritated. In fact, in a letter that Selberg wrote in September 1948 to Hermann Weyl, an influential member of the Institute for Advanced Studies, one can read the following:

> I didn't like it at all that someone else started working on things I hadn't published yet. Although I was very irritated by this, I did not say anything to Erdös, because it seemed to me that, after all, he was interested in things other than mine. However, I seriously started to worry, and so I decided to work very hard on my own ideas. On Friday evening on July 16th, Erdös told me that he had succeeded in proving his formula of the ratio. On Sunday, I had my first proof of the prime number theorem and a few days later, I got another one, which was much more satisfying.

Meanwhile, Erdös had also arrived at the elementary proof of the prime number theorem, and he therefore proposed to Selberg that they write a series of papers to be submitted to the *Annals of Mathematics*, one of the most prestigious scientific journals, according to the following scheme: first, an article by Selberg on his fundamental formula, second, another article, signed by both, in which the formula of the ratio would be initially introduced, followed by the proof of the prime number theorem obtained in an elementary way. Meanwhile, as was his custom every time he had conquered a new summit of mathematics, Erdös sent dozens of postcards to his many collaborators around the world with the following text:

> Using a fundamental formula by Atle Selberg, Selberg and I were able to give an elementary proof of the prime number theorem!

From that moment on, however, things began to deteriorate, going from bad to worse. The reason for this was that Erdös was already well known, while Selberg was unknown to most. So, when, shortly thereafter, Selberg visited the mathematics department at Syracuse, he was asked by its director, "Did you know that Erdös and another guy had obtained an elementary proof of the prime number theorem?"

Selberg began to fume with rage, and he made clear his anger in another letter to Hermann Weyl:

> In Syracuse, I discovered from several people that there was great publicity about a new proof of the prime number theorem. All of the people to whom I talked attributed the entire proof to Erdös, even those who knew my name and my previous work well.

When he returned to Princeton, the story immediately took another turn. Selberg coldly rejected all of Erdös's attempts to publish a jointly signed paper on that great discovery, and, in his letter to Erdös of September 20, 1948, he made his feelings clear:

> I hope that we can reach an agreement, but I will never accept publishing an article together on this problem.

The reason, later explained to a friend, was as follows:

> Erdös replied to me that we had to do it like Hardy and Littlewood. But we had no agreement! In fact, we never had any collaboration. His involvement in this subject was entirely casual, it was never my intention to involve him.

Erdös's reply arrived a few days later, on September 27, 1948:

> I completely reject the idea of publishing only the formula of the ratio, and I am instead more and more convinced that we should publish a paper together. If, instead, you insist on publishing the paper on your own, what I will do will be to publish our common proof, giving you full credit for your contribution, i.e., that you obtained the prime number theorem first, but using some of my ideas and my own formula of the ratio.

There continued to be a great chill between the two. Selberg wrote his article, under his own name, in which his elementary derivation of the prime number theorem was presented: the article, in which Erdös was only mentioned en passant, was immediately published in the *Annals of Mathematics*. Erdös also wrote his article, giving ample credit to Selberg for his elementary formula, from which he derived—he wrote in the text—the formula of the ratio, and from which the prime number theorem ultimately followed. It was sent to the same journal, but the article was rejected, owing to the direct pressure of Hermann Weyl, as it was discovered later. Erdös's article was finally published, with a delay of several months, in another journal.

The aftermath of this querelle poisoned the community of mathematicians for years, splitting it between those who sided with one and those who sided with the other. There were those who, like Erdös—a mathematical Don Juan who was always ready to conquer every fine problem that he came into contact with, and who had made mathematics into a joyously social activity, extended to tens or even hundreds of people—found it utterly difficult to understand the reluctance and the coldness of someone like Selberg. Vice versa, for those who were like Selberg—a Sir Galahad, completely devoted to the purity and nobility of the soul, a Magister Ludi from the realm of mathematics, who saw it as a purely monastic activity to be pursued in solitude—understanding the exuberance and the intrusiveness, no matter how good-natured and involuntary, of someone like Erdös was an equally impossible undertaking. The two were simply destined never to understand each other, and, after the events of that summer of 1949, their paths were, in fact, divided forever.

Further Readings

M. Aigner, G.M. Ziegler, *Proofs from the book* (Springer, Berlin, 2006)
P. Hoffman, *The Man Who Loved Only Numbers* (Hyperion Books, 1998)

Faraday. An Electric Discovery

Friday Evening at the Royal Institution

On the evening of Friday, June 20, 1862, a large crowd had crept into a street in the heart of the West End of London, a turbulent river of cylinder hats and pretty lace caps. Among the wide folds of the crinolines, there were the faces of several children, also excited by all of that turmoil and the strange agitation seemed to possess their parents. And the tension, as the minutes passed by, did not seem to be diminishing, quite the contrary. At the corner of the road, the clatter of the horses and the coming and going of the carriages continued incessantly. Elegant ladies and distinguished gentlemen rushed out of the little doors of the carriages and ran to make an even longer queue at the entrance to the majestic colonnade of the building located on 21 Albemarle Street, home of the Royal Institution. It seemed like the hustle and bustle of a theatrical premiere or a famous opera, in any case, an event that nobody wanted to miss. It was, indeed, a very special day, but for an unusual reason: that evening, the curtain would come down on extraordinary performance, a show that had lasted continuously for almost thirty years. It was the evening of the moving farewell of a scientist, now old and sick, in front of the public that had admired him for years. A great scientist, but also a brilliant storyteller, who had delighted the young and the old every Friday, with fascinating explanations of natural phenomena: using just simple words, he showed them the surprises that were normally reserved for the fields of chemistry and electromagnetism. With great skill, he had magically unfolded the fantastic book of nature before their eyes.

In the audience of that last evening, there were those who still remembered the amazing experiment of the metal cage, a device able to completely shield anything inside of it from external electric fields. One evening, he placed a cat in the cage, and the little animal remained lying inside, relaxed and blissful, despite the sparks and electric shocks from thousands of volts that spilled outward from the walls of the cage. There were also those who still had a clear memory of the magnetic needle that moved like crazy when an electric coil was turned on. And these were not the only wonderful things seen during those years! Every Friday evening, the semicircular

© Springer Nature Switzerland AG 2020
G. Mussardo, *The ABC's of Science*,
https://doi.org/10.1007/978-3-030-55169-8_6

hall of the Royal Institution was always packed with people. But the lucky ones who could find a seat were sure to be in for a sensational show: to see, for instance, the electrolytic decomposition of the water in the gas bubbles of oxygen and hydrogen, to look with wonder at the blue color that the gases assumed once liquefied, to be able to witness in awe the lines of force that iron filings traced around a magnet, or to remain completely fascinated by the enchanting explanations as to what makes the flame of a candle agitate.

In the laboratories below the hall, in the building that had been his entire world for almost fifty years, that scientist had conducted deep investigations into the relationships that link electricity to magnetism. He had managed to prove a sensational fact: that an electric current could be induced in a circuit using a variable magnetic field, a truly electrifying discovery, which had brought him honour and fame all over the world. In those laboratories, he had conducted research on diamagnetic and paramagnetic substances. He had then cleverly exploited the ferromagnetic ones to artificially create a magnet and build the first electric motor. We owe an impressive series of discoveries to that man, discoveries that were destined to radically change the world and to pave the way for new and profound scientific and technological revolutions. The scientist upon whom the curtain was closing that evening was Michael Faraday.

A Dickens Character

Each era has its own symbol characters, those personalities who are able to magically evoke the full atmosphere of a period with their names. Just as Elizabethan England is beautifully summed up by the name of Shakespeare, the Victorian one is synthesized—for almost antithetical reasons—by the names of Charles Dickens and Michael Faraday. With Dickens, the art of literature of the nineteenth century reached its peak, mixing crazy laughter with an almost unconscious descent into darkness. Faraday, contrastingly, wrote one of the most beautiful pages of experimental physics. Dickens—with the muddy water of the Thames, the yellow-dark fog that covered London like a sooty veil and an endless and desperate series of dropouts—represents the darkest image of the Industrial Revolution and England of the 1800s. Faraday, with his amperometers, magnets and test tubes, is, instead, its brightest icon. If Oliver Twist, the Artful Dodger or David Copperfield—extraordinary characters from Dickensian tales—immediately remind us of the crude reality of the exploitation of child labour, family violence or the miserable living conditions in London slums, a character such as Faraday still represents to us today the enthusiasm of the great discoveries and the vibrant scientific atmosphere of Victorian London. Moreover, the story of his life is unbelievable: the moral of it seems to be that the things that really matter are tenacity and passion, and that true passions always find their fulfilment, in spite of all of the worst adversities. Michael Faraday indeed had something extraordinary, being, at the same time, the perfect incarnation of a Dickensian character and the protagonist of a story with a happy ending.

Michael Faraday was born on September 22, 1791, in a slum suburb in south London. His was a humble and cultureless family. His father, a farrier, could barely support his wife Margaret and their three children. Although desperately poor, the Faradays were, however, an unusually happy family. The great strength behind them came from their religious faith, as they were members of Robert Sandeman's fundamentalist sect, based on a literal interpretation of the Bible, and on unshakable trust in the heavenly salvation that awaited his followers, a certainty that helped them to tolerate all of the hardships of everyday life. The Sandemans had little interest in material goods, but an unconditional love for their neighbours and a strong sense of brotherhood, similar to that which animated the ancient Christian communities. This strong religious imprint on the family remained a dominant feature of Michael Faraday's personality throughout his life: in addition to a strict moral code that caused him to refuse many honours, it also gave him the serenity necessary to face the heaviest vicissitudes. Faith also helped him in his work as a scientist: he was convinced that understanding the secrets of the Book of Nature written by God was as important as reading his other great text, the Bible.

As a boy, Michael Faraday received a rather rudimentary education. His scholastic experience ended abruptly, due to his lack of facility for pronunciation: he was unable to articulate the letter R well, and his acidic teacher had repeatedly humiliated him for this. In addition to making him the laughingstock of the class, she also at least once tried to bully him physically: she summoned a pupil and gave him half a penny to go buy a stick with which she intended to beat Michael. The classmate, however, went to Faraday's house and reported this teacher's cruelty to his mother. Mrs. Faraday rushed to the classroom and withdrew Michael from school: the health of her son was much more important than his education.

From then on, Michael spent all of his time on the street, playing with stones and "watching the sunsets," because, as he remembered years later, "a beautiful sunset always brought thousands of thoughts that delighted me." His was a real struggle for survival: in times of famine, the only sustenance was a meager piece of bread that had to last a whole week. His adolescence ended abruptly at the age of thirteen, when, due to his father's precarious health, he was forced to go to work at a bookstore owned by a French refugee named George Riebau.

The Riebau Bookshop

Despite appearances, this was his first stroke of luck. London bookbinders, in the turbulent years of the early nineteenth century, were very interesting places: they were not only shops where books were bound and sold, but also theaters of lively discussion and extravagant encounters. The one located at No. 2 Blandford Street was a fairly good example of this and Riebau was a good man too. At the beginning, Faraday was assigned to the distribution of newspapers and the recovery of unsold copies, but since Riebau was very fond of the boy, after a year, he decided that it was appropriate to teach him the profession of bookbinder and took him on as an apprentice. Faraday was very talented and showed remarkable skill: it was a pleasure to see how he mastered the large pages full of ink, to observe how he carefully folded and sewed the edges, to admire how the books finally took shape in his hands. Besides binding them, however, Faraday also read the books, and voraciously at that! His was an enthusiastic and indiscriminate reader: "Do not think I was very deep at the time," he told John Tyndall many years later, his friend and successor at the Royal Institution, "I had only a great fantasy. I could easily believe in the Thousand and One Nights as much as the Encyclopaedia Britannica. The only things that saved me were the facts, these were very important to me. I could believe a fact only after I had scrutinized it carefully and verified it personally." For Faraday, the "facts" were as sacred as the verses of the Bible, so, in the evening, when the laboratory was finally empty, he would try to repeat the experiments reported in the books that he had read.

Electricity and Chemistry

In particular, he was galvanized by reading a volume of the Encyclopaedia Britannica from which he learned of the existence of the electric phenomena that was arousing so much curiosity at that time: the experiments done with lightning by Benjamin Franklin, not to mention all of the stories about the powerful electric discharges that manifested upon touching the famous bottle of Leiden, a large glass vessel lined inside and out with a sheet of foil. The text of the Encyclopaedia Britannica had been written by a certain James Tytler, and his point of view, compared to the opinions of the time, was quite original. According to Tytler, in fact, all of the electrical effects could be explained on the basis of the existence of a fluid, whose vibrations accounted not only for the galvanic currents and the functioning of the Volta battery, but also for optical and thermal phenomena. It is difficult to say what Tytler's influence was on Faraday's future ideas about such a fundamental concept as that of the lines of force of the electric and magnetic field: the fact remains, however, that continuous references to Tytler's volume can be found in the diary that Faraday began to write in those years, pages in which he kept note of his ideas and the first experiments that he carried out in the back of the typography.

Another great discovery was the book by Jane Marcet, *Conversations on Chemistry*, written in the form of a dialogue, whose protagonists—a housekeeper and two maids—drew his attention for the first time to the chemical philosophy of Humphry Davy, the great English chemist who would play a decisive role in his life. For Davy, chemistry represented the key to understanding the mysteries of nature, which is why the book did not end up being a classically arid catalog of facts, but rather a true scientific fresco in which chemical reactions, electrical relationships, and optical and thermal phenomena took place simultaneously. The impact on Faraday was noteworthy: he temporarily put aside his interest in electricity, and instead directed his attention overwhelmingly towards chemistry. Humphrey Davy soon became his idol, the man he wanted to be.

Public Lessons

However, there was a problem: science in the first half of the 19th century was in a transitional phase. Translated in practice, in those years, there was not a definite path, or a consolidated practice, to becoming a scientist: the term "scientist," for example, did not exist, first being coined in 1833. There were no university diplomas, much less PhD courses to follow. The few places available to engage in science were usually the prerogative of those wealthy gentlemen who, having the necessary time and money, could devote themselves without problems to the speculations of natural philosophy. It was also a matter of discussion as to whether science was a vocation or a career, a serious profession or entertainment. In fact, in Victorian London, various and renowned scientific institutions flourished, the most famous of which was the Royal Society.

There was also a long tradition of public lessons, albeit ones that had to be paid for, an apparently very lucrative activity for the lecturers, given the great and general curiosity in new discoveries. The lessons that dealt with medicine, chemistry, geology and mineralogy were especially popular. There were also large crowds for topics on astronomy and electrical experiments, the latter definitely being a goldmine for those who wanted to stage the extraordinary show offered by natural forces. People were thrilled to see the frogs' legs twitching under the action of an electrode, or for a whole series of new crazy things, such as the electric shocks that could be felt simultaneously by a chain of people, or seeing little boys who, suspended in the air with insulating silk cords, began to emit sparks from their hands and feet, with their hair all standing on end, as soon as they were connected to electrostatic machines.

A Lucky Encounter

If the meeting with Riebau at the age of thirteen was the first stroke of luck for Faraday, the second happened to him when was eighteen years old: it was when he read, by chance, the announcement in a newspaper of a series of public lessons in natural philosophy, a term then used to generically indicate science, at the City Philosophical Society. The inscription cost was one shilling per lesson, not a negligible sum, considering his modest finances. His older brother came to his aid, generously taking charge of the fare. Faraday attended the lessons with a growing feeling of dizziness, carefully writing down everything that he heard and saw: reasoning, demonstrations, settings, reports of experiments. He then collected these notes in a series of volumes. Binding, after all, was his job!

In spite of the euphoric enthusiasm with which he threw himself into all of these novelties, it seemed that there was very little possibility that Michael Faraday, apprentice bookbinder, would become anything other than Michael Faraday, bookbinder. But sometimes fate helps the talented: the four large volumes of notes that he collected were proudly shown by Riebau to one of the most assiduous and extravagant customers of his library, a certain Mr. Dance, also quite passionate about science. Struck by the richly decorated books, Mr. Dance asked if he could borrow the books in order to view their contents and also show them to his father.

The result of these happy coincidences was that the elder Mr. Dance was extremely pleased to provide the young apprentice with tickets to attend the public lessons of his hero, the famous English chemist Humphry Davy, director of the Royal Institution. The choice by Count Rumford—the founder of this institution—to place Davy as the first director of the Royal Institution had proved very apt: the brilliant oratory of the young chemist ensured both the success of the lessons and the full house. Davy was also a very good-looking man, a detail to which the young women of high society seemed to be very sensitive, with legions of them filling the audience at every show. For Faraday, to participate in Davy's lessons was like touching the sky with his finger: in those hours, Davy spoke about the transmutation of matter, the laws of gases, galvanic currents, chlorine and carbon compounds, the discovery of oxygen and its explosive combustion, and the extraordinary things that happened when different substances were combined. The contents of these lessons were first transcribed by Faraday in beautiful calligraphy, and then printed with the matrices of the Riebau typography. He sent the volume to the illustrious chemist and, as happened previously with Mr. Dance, Davy was so impressed that he immediately summoned the boy to fill the position of his assistant. This was the stroke of luck that definitively paved Faraday's way to the path of science. The apprentice bookbinder would no longer deal with the books of others; it was his turn now to write some of the most beautiful pages in the history of physics.

"In the Future, You Will Put a Tax on It"

Michael Faraday was hired by Humphrey Davy as his assistant at the Royal Institution on March 1, 1813. The following year, he was invited to accompany Davy and his wife, a wealthy and arrogant aristocrat, on a long journey through Europe. Although Mrs. Davy did everything to mock the young man and remind him of his humble origins, it was a fundamental journey for Faraday's education: he got to know the most influential scientists of the time, such as Ampère, Volta, Arago and Gay-Lussac, with whom he kept in contact later.

When he returned to England, he began his extraordinary scientific career. In the early days, Faraday helped Davy conduct his chemistry experiments, a discipline that had always fascinated him. Working in the field of electrochemistry—inaugurated by his master Davy, who had discovered a whole series of new elements, such as potassium, boron, and iodine—Faraday introduced a new terminology, which has remained in use since then, and an important series of new physical laws. He also identified benzene and ventured into the liquefaction of several gases.

He began to deal with electromagnetism after the famous observation made in 1820 by the Danish physicist Hans Christian Ørsted: during a demonstration of the transport of electric current in a metal wire, he had noticed that a magnetic needle, left nearby, was deflected violently every time he turned on the circuit! This meant that any variation of the electric field was able to create a magnetic field that influences the direction of the needle. Faraday was very impressed: one year later [?], he had managed to invent a small electric motor, a device capable of continuously rotating a magnet around the axis through which the current flowed.

Before Ørsted, in the field of electromagnetism, there had been the experiments systematically carried out by Francois Arago and André-Marie Ampère: the latter, in particular, was able to show that an electric current flowing through a wire twisted around a coil made the whole device equivalent to a small magnet. With a flash of genius, Faraday repeated the experiment by placing two coils, one in front of the other: he noticed that, by passing the current through only one of the two, thanks to a battery, a current was also induced in the other, as could be seen from the deflection of a magnetic needle placed in its center: in a few words, after 10 years of hard work, he had come to the fundamental law of electromagnetic induction. If Ørsted had shown that a magnetic effect was induced by an electric field, Faraday managed to prove the opposite, that is, that a magnetic field could induce an electric field. This discovery transformed studies on electricity and magnetism from an object of pure curiosity into a new powerful technology. William Gladstone, Finance Minister and Chancellor of the Exchequer of Queen Victoria, once visited Faraday in his laboratory. As he saw the strange devices scattered on the table, Gladstone dropped the question, "Interesting, but what is its practical use?", to which Faraday replied laconically, "At the moment, I don't know, Sir, but it is very probable that, in the future, you will put a tax on it."

A Choice of Field

Faraday later made another fundamental observation: in letting light pass through a strong magnetic field, he noticed that its polarization was altered. There also had to be a very deep connection between electromagnetism and light! From all of his experiences, he had learned one fundamental thing: all of the forces of nature were related to each other. In one of his lectures at the Royal Institution, he said, "I am of the opinion that all the various forms in which different forces manifest themselves actually have a single origin: they are mutually dependent and convertible into one another, and furthermore they have the same power."

Among the various cunning methods that he invented to study electro-magnetism, there was one of which he was particularly proud, including for its "philosophical" consequences, even though that is quite an ironic statement for a person as pragmatic as Faraday. The method was to spread iron filings on a sheet of paper and see how the small metal flakes became oriented once subjected to a magnetic field: the result was a wonderful representation of how space was influenced, point by point, by the presence of the magnetic field. Lines of force—as they were called by Faraday— were of a higher number where the magnetic force was stronger: as a fluid, they seemed to completely fill the space. He then assumed that those lines of force existed regardless of whether he highlighted them with iron filings or not. Each system of charges or currents—he very much liked to think—creates an electromagnetic field, a disturbance of space that can vary if the arrangement of the charges or the intensity of the currents varies. The same had to happen with the gravitational field, created by the presence of bodies with mass.

Faraday, in particular, did not believe in the existence of an action at a distance between the bodies. Each force had to be mediated by a field and could never be instantly transmitted remotely: the field is the only object that has physical reality, being that it occupies space, brings energy and eliminates the void by its presence. Only a few people understood this idea. William Thomson of Glasgow was the only one able to appreciate the strength of this concept. "I am happy to wait, convinced of the truth of my ideas, someone else will follow them," said Faraday. Prophetic words, indeed: in 1861, James Clerk Maxwell founded the theory of electromagnetism starting from Faraday's field concept, forever changing the course of physics.

Further Readings

E. Segrè, *Great Men of Physics* (Tinnon Brown, 1969)
E.T. Canby, *The History of Electricity* (Hawthorn Books, 1969)

Germain. Sophie's Choice

The history of science is full of episodes of discrimination against women: one of the most striking episodes remains that of Lise Meiter, the subject of one of our future chapters. Sometimes, it was the malice of words that took over, as in the case of the mathematician Maria Gaetana Agnesi, who lived in the second half of the eighteenth century and was known for an algebraic curve called "the witch of Agnesi": this name came from strange alliterations of the original name, but it says a lot about the connotative power of words and what they can insinuate. But now, we will look at the story of woman who found her own unique way to make her voice heard by the mathematical minds of her day.

Monsieur LeBlanc

It may seem strange that such discrimination would happen within the scientific community. It is commonly believed that science is a discipline so ethically sublime as to give everyone the right to speak, regardless of wealth, "race," gender or age, in which the only thing that matters is the logical strength of the arguments. This is, unfortunately, too naive of a consideration: the daily management of science is, in fact, under the direct responsibility of those who practice it and, as such, is subject both to prejudice and to relations of academic power, both very difficult to unhinge. Perhaps a greater female presence would help science to fly towards a much higher standard, as evidenced by the emblematic story that we are going to tell.

The history of this great scientist recalls moving episodes such as those described by Charles Dickens, as well as surprising coups de théâtre worthy of Alexandre Dumas's pen: in order that her discoveries might be known, our brilliant mathematician decided to disguise herself as a man. The following is the story of a surprising change of identity that lasted for years: in short, it is the story of Sophie Germain, also known as "Monsieur LeBlanc."

© Springer Nature Switzerland AG 2020
G. Mussardo, *The ABC's of Science*,
https://doi.org/10.1007/978-3-030-55169-8_7

The Tragic End of Archimedes

Sophie Germain was born in Paris on April 1, 1776, the second of three daughters of a wealthy family. Her father, Ambroise-François, a textile merchant, later actively entered the political life of France, participating first in the Assembly of General States, and then in the work of the Constituent Assembly. Germain's youthful years coincided with the turbulent moment of the French Revolution. At that time, the streets of Paris were full of excited people, armed with forks, scythes and sticks: clashes with the king's troops were always on the agenda, and there was a daily risk of a bloodbath. With the arrest and subsequent beheading of King Louis XVI and Queen Marie Antoinette, the period of the Terror started, and the guillotine began to work at full speed in the squares of Paris. Ambroise Germain decided that it was wiser to take the family out of town so as to isolate them from the chaos that reigned in the streets of the capital.

They moved from their apartment on Rue Sainte-Croix de la Bretonnerie, near Notre Dame, to a country house. There, to find comfort in the boredom of the days spent without friends, Sophie immersed herself in reading the numerous volumes that filled the shelves of her father's huge library. There were, in particular, two that caught her attention: the first was a text written by the French mathematician Étienne Bézout, a very popular book used both in French schools and by scholars of the time; the second, *Histoire des Mathématiques* by Jean-Étienne Montucla, was instead a volume the subject of which was the ancient Greek geometers, the Italian algebraists, and the seventeenth-century English mathematicians, perhaps the first known book of the true history of mathematics. Among the various events narrated by Montucla, the one that struck Germain most, at the age of thirteen, was the tragic end of Archimedes of Syracuse, killed for not paying attention to the yelling and threats of a Roman soldier because he was immersed in the solution of a geometry problem. Sophie was so impressed that she decided to dedicate her life, from that moment onwards, to mathematics: that world made of numbers, formulae, geometric figures, subtle reasonings, had to be a truly fascinating land if a great thinker like Archimedes had lost his life for it!

The Forbidden Texts of the Great Mathematicians

Germain was overwhelmed by a dazzling passion, with all that follows: her love for mathematics pushed her to study Latin and Greek as an autodidactic, in order to be able to read the classic texts in the original language. Her enthusiasm, however, was not at all approved of by her parents, who tried all means to discourage her from continuing with those perverse readings: they prohibited her from using candles, and even went so far as to prohibit the lighting of the fireplace in her room to prevent her from reading in the evening. All of this was useless: she made up for the candles seized with ones that she made herself; she hid them under her skirts and took them

to her room, where she used to spend the nights wrapped in a blanket to protect herself from the cold of the extinguished fireplace, intent on reading the "forbidden texts" of the great mathematicians of the past.

The faint light of those artisan candles was, however, sufficient to open up radiant horizons before her: if the pages of Newton revealed the mysteries of the laws of motion, those of Laplace's celestial mechanics led her to discover the majesty of the cosmos and the elegance of mathematics, which was revealed to her as a very powerful language for describing in detail the gears of that enormous celestial mechanism; Euler's writings, rich in extraordinary formulae, caused her to discover the vertigo of infinity, while those of Gauss led her to the wonderful world of number theory. Each day was one of discovery, an emotion incomparably greater than that experienced in playing with her peers or in the daily life of receptions, lunches, courtesy visits and sterile parlor conversations.

The account of the Dickensian atmosphere breathed in her youth by Sophie Germain came to us thanks to the direct testimony of Guglielmo Libri Carucci, a singular character of the academic world of the early nineteenth century, an Italian mathematician well integrated into various French intellectual circles. Even re-dimensioning its tones, certainly, for that Parisian family of solid bourgeois roots to have a daughter devoted all day to reading foreign texts full of formulae and geometric designs, showing total intolerance towards dancing, afternoon tea, walks with friends and all of the good opportunities to find a husband, had to be really strange.

Germain was very shy, but hard as a rock: in the end, her determination broke through, and her parents allowed her to devote herself to that which she held most precious—her studies, her mathematical scribbles—without hindering her, ultimately even supporting her economically for a lifetime.

A Mysterious Pupil

If her parents gave up, the academic world of the early nineteenth century didn't. That a woman could have a passion for mathematics was absurd, to say the least. Enroll in university and attend courses? Simply impossible. But Germain had now been vaccinated against this strange disease of bans: as a child, she had had to smuggle candles; now she would do the same with lecture notes. She borrowed them from some student at the Ecole Polytechnique, a friend of hers, and returned them after studying them avidly. At the time, it was customary for students to send comments on the topics of the courses or the problems assigned by the professors. Various teachers thus began regularly to receive a singular stream of missives containing witty reasonings on advanced mathematical topics. The letters were signed by Monsieur Antoine-Auguste LeBlanc, a mysterious pupil whom no one could identify in his class.

Among the professors, there was also the famous mathematician Joseph-Louis Lagrange, who, amazed by the acumen of these notes, wanted to know this talented student. His amazement was significant when he discovered that a woman was hidden behind that name! A certain turmoil ensued, the news passing from mouth to mouth, from class to class, from living room to living room: it was certainly not every day that a woman was so passionate about mathematics. Hence, curiosity mounted, and many scholars came forward to meet "Monsieur LeBlanc."

Not all, however, were congenial encounters. One of these involved the astronomer Jérôme Lalande, director of the Paris Observatory, as well as author of the most complete star catalogue of the time. When he met Sophie Germain, he was a sixty-five year old man, a regular guest of many lounges on the Parisian Rive Droite, where he loved to entertain the ladies with his jokes and stories. His fault was mistaking Sophie for one of the typical ladies to whom he was accustomed: when she tried to tell him that, among her scientific readings, she had devoured Laplace's *Système du Monde*, Lalande interrupted her by saying, with arrogance, that to understand that book, one had better first read the volume that he had written on astronomy, a volume certainly more suitable for women than Laplace's. This book was titled *Astronomie des Dames* and followed the lines of other famous texts of the time, such as *Newtonianism for Ladies*, written by the Italian scholar Francesco Algarotti, under the assumption that women were only interested in courtship and love, and that the only thing they wanted was a smattering of science with which to show off when, between the velvets and the armchairs of the Parisian lounges, they were grappling with the subtle art of conversation. At the heart of this type of book, there was always a certain cliché, according to which a man—typically a marquis or a high-ranking gentleman—conversed with a lady to whom he explained the various physical laws with analogies and simplifications: the fact, for example, that the gravitational force between two bodies decreases with the square of the distance was expressed by asserting that the same thing also happens in the sentimental field, because four days of distance make love sixteen times more feeble… To Lalande's words, Sophie Germain reacted not only with indignation, but great anger! Before turning back abruptly, she screamed in his face that she did not need his text at all, that she was not like Madame Leroy or the Duchess of Guermantes, but was instead genuinely interested in science, not in the art of conversation. The following day, the astonished astronomer sent her an apology letter in which he wrote:

> Mademoiselle, never, before yesterday, was there anyone who made me feel so imprudent about my visit and the disapproval of my respect. But I must say that it was very difficult to predict such a development. […] You told me you have read Laplace's book and that you did not wish to read my book on astronomy. I remember saying that one cannot understand the former without reading the latter. It must have been this comment of mine that made you so angry. I deeply apologize, Lalande.

Not only did Germain not accept the apology, she also made sure to not meet Lalande again, anywhere or under any circumstances. Fortunately, with other scholars, it went better; among these was Adrien-Marie Legendre, a very famous mathematician and author of several treatises, including one dedicated to number theory. Perhaps it was as a result of Legendre's direct influence—who knows?—that Sophie Germain became so passionate about this field,

Give the Numbers

Number theory is one of the oldest and most fascinating branches of mathematics, born from questions about prime numbers posed by the ancient Greeks. Is there a more immediate concept than prime numbers? They are natural numbers divisible only by 1 and themselves: apart from 1, the sequence opens with 2, 3, 5, 7, 11, 13, 17, 19, 23... Since any other integer can be uniquely written as a product of primes—a fundamental result of mathematics—prime numbers can be considered the atoms of arithmetic. However, unlike the chemical elements, which are finite in number, the prime numbers are infinite. The proof—extraordinarily elegant in its simplicity—dates back to 300 B.C. and is credited to Euclid: if the primes were finite and in number k, reasoned the great Greek mathematician, it would be possible to construct a natural number N by making the product of all of the primes and adding 1 to the end:

$$N = p_1 p_2 \cdots p_k + 1.$$

Clearly, this number is not divisible by any of the prime numbers up to p_k, so either N is a new prime number, larger than the previous ones, or it can be factored in terms of other primes larger than those initially considered. Chapeau!

Among the founding fathers of modern number theory, Pierre de Fermat, a professional judge, should also be mentioned. Born near Toulouse in 1601, he spent most of his life in southern France, away from the major centers of knowledge of the time, practicing the forensic profession full-time and only dedicating his spare time to the art of mathematics, of which he was the prince of amateurs! However, his studies had a huge impact on the development of the discipline. His theorems circulated thanks to the dense correspondence that he undertook with Father Marin Mersenne and with all of the most important mathematicians of that time, although the fact that he did not often reveal the proof of his theorems was a cause of great frustration: the Englishman John Wallis actually called him "that damned Frenchman," while his compatriot Descartes considered him a "braggart." Together with Blaise Pascal, Fermat developed the theory of probability; he also dealt with problems concerning the minimum and maximum of functions, even before differential calculus was developed. In the field of number theory, Fermat found his main source of inspiration in a treatise on the great Greek mathematician Diophantus of Alexandria, *Arithmetic*. This text had only emerged from the oblivion of time in the seventeenth century, and Fermat, who owned a copy, held it in high esteem: not only did he read and reread it, but he used the margins to take note of his progress and the new results that he was gradually discovering. And, as it turned out, a simple side note ended up becoming one of the most famous questions in the history of mathematics, remaining open for centuries and not being definitively resolved until 1994.

The problem that we are talking about originated in the Pythagorean theorem, the famous relationship that links the square built on the hypotenuse of a right triangle to the squares built on the two cathets. Indicating the two cathets with a and b, and the hypotenuse with c, we have the familiar equation

$$a^2 + b^2 = c^2.$$

So far, no surprises. However, it starts to get interesting if the three numbers a, b, c that satisfy the Pythagorean equation are required to be integers: in restricting the field to integers only, is there still a solution to this equation? This is the typical example of the problems faced by Diophantus (finding solutions to algebraic equations in terms of integers) and similar equations are thus called "diophantine equations." For Pythagoras' theorem, it is easy to prove that there are infinite terns of integers that satisfy the equation, known as "Pythagorean terns": the simplest example is given by (a, b, c) = (3, 4, 5), followed by (5, 12, 13), and so on.

In light of this result, it was legitimate to ask, as Fermat did, whether generalizing the problem, as well as the equation

$$a^n + b^n = c^n,$$

had integer solutions in terms of the three numbers a, b and c in the event that the exponent n was greater than 2. On this question, Fermat took care to leave an enigmatic note in the margin of the Diophantus text that he always carried with him: "It is impossible," he noted, "to write a cube as the sum of two cubes or a fourth power as the sum of two fourth powers or, in general, no number that is a power greater than two can be written as the sum of two powers of the same value. I have a wonderful demonstration of this theorem which cannot be contained in the two narrow margins of the page." This statement went down in history as *Fermat's last theorem*, although there is no trace among his papers of the alleged general demonstration that he claimed. The only particular case proved by Fermat was, in fact, the one related to n = 4, for which he invented a new method of mathematical demonstration, known as the method of infinite descent. It is a variation of the classic *reductio ad absurdum* proof and consists in setting the proof of a theorem involving natural numbers on the existence of a set that enjoys a certain property; there must necessarily be a minimum of this set, and if, once you have taken this smaller number of the set, you can prove that there is an even smaller one, then you have fallen into contradiction. It is a very cunning procedure, but one that, however, prevented Fermat from going beyond the case of n = 4.

Correspondence with the Prince of Mathematicians

When Sophie Germain became interested in this problem, almost two hundred years had passed since its formulation, and the only progress made in the meantime was credited to Euler, who, in 1770, had managed to prove the case of n = 3. The rest was earth unknown. In 1801, Gauss, the prince of mathematicians, had published the volume *Disquisitionses Arithmeticae*, which was immediately accepted as the key text of number theory, a discipline finally presented organically and not as a simple collection of sparse problems. Germain sank into the reading of this magnum opus and quickly learned to work with modular relations and the divisibility properties of numbers, the focal points of Gauss's treatise. These served her in making a significant step forward regarding Fermat's last theorem: in fact, she managed to demonstrate that it does not admit a solution for a series of values of the exponent n linked to particular prime numbers, now known as Germain's prime numbers, destined to remain the most important result on the subject for a long time.

Germain felt compelled to seek the approval of the Principal of Mathematicians, to whom she wrote for the first time on November 21, 1804, signing the letter, as usual, as Monsieur LeBlanc. Although Gauss was not very interested in Fermat's theorem—he had previously written to his astronomer friend Heinrich Olbers, "Fermat's last theorem interests me very little, since I could easily find a multitude of similar propositions, which could be neither proven nor refuted"—he replied politely, saying that he had read with pleasure what Monsieur LeBlanc had written and was happy with this new friend, known through their common passion for mathematics. So encouraged, Germain wrote to him again, the correspondence between the two thus lasting several years. A letter from the great Gauss followed another from Monsieur LeBlanc, and the true identity of the latter would have remained unknown forever if Napoleon had not invaded Prussia in 1806. When, after the victory at Jena, the French troops moved towards Brunswick, where Gauss lived, Sophie Germain was very worried that he would fall prey to the same tragic end as Archimedes. She thus wrote to a family friend, General Joseph-Marie de Pernety, commander of the French artillery in the Prussian countryside, urging him to make sure that the German mathematician was in no danger. When the general was able to trace Gauss and told him that this gesture of respect for him was due to the interest of a young Parisian mathematician, Sophie Germain, it was if the name had fallen from the clouds, unfamiliar as it was to Gauss. The mystery was cleared up with the arrival of a new letter from Paris:

> General Pernety informed me that he made my name to you. This leads me to confess that I am not as unknown to you as you may think, because, fearing the sense of ridicule associated with a female scientist, I previously preferred to adopt the name of Monsieur LeBlanc to communicate to you those notes that you had indulged me in reading...

Gauss replied immediately, with amazed words of admiration:

> When a female person who, according to our male judgments, has to run into difficulties infinitely greater than those encountered by men to get to know the thorny research of mathematics, when this person manages, despite everything, to overcome such obstacles and penetrate the darkest regions of science, she must undoubtedly possess a noble courage, an absolutely extraordinary talent and a superior genius.

Certainly the most beautiful homage that could be paid to Germain's intelligence.

Sophie Germain, and Monsieur LeBlanc with her, died in 1831. Going by the notice of her death it appears that she was neither a mathematician nor a scientist: in the eyes of her contemporaries, she was only a simple "landowner."

Further Readings

A. Dahan Dalmedico, Sophie Germain Scient. Am. **6**(256), 116–123 (1991)

S. Singh, *Fermat's Last Theorem* (Fourth Estate, London, 2002)

Harriot. Down the Spiral

The history of mathematics has lost track of Thomas Harriot. There is no well-known burial place or monument to remember him, as with Newton, buried in the Westminster Abbey with full honours. There is no formula or equation that bears his name: during his life, he published a single publication, dedicated to the study of the language and customs of American Indians, while the rest of his vast scientific production was lost in boxes full of dusty manuscripts. Moreover, the posthumous publication of this impressive notebook had an incredibly tangled story—a true bibliographic thriller, with false twists and turns, manipulated formulae and counterfeit texts, printed mostly by individuals in search of fame. Orienting ourselves in this labyrinth is not an easy task, and it is only recently, thanks to the meticulous work of several science historians, that the importance of this scholar has finally emerged, recognized today as the greatest English scientist before the advent of Newton. His is the portrait of a brilliant character, a brilliant and acute pioneer of several branches of science, who lived in the turbulent period of Sir Walter Raleigh and Christopher Marlowe, at the Elizabethan court in 16th century England, a world full of intrigue, sudden fortune and turbulent overturns. Hence, who was Thomas Harriot?

In Elizabethan London

The first traces that we have of Thomas Harriot date back to his enrollment at the University of Oxford, certified by a document dated 1580, which has allowed us to estimate his date of birth at around 1560. At Oxford, he obtained a diploma—typical for the time—in the study of the Greek and Latin classics, but his real interests were already turned then towards astronomy and geometry. Moreover, he had a real passion for cartography: he was fascinated by the dizzying sequence of geographical discoveries that followed in those years and by the wonderful maps drawn up in

© Springer Nature Switzerland AG 2020
G. Mussardo, *The ABC's of Science*,
https://doi.org/10.1007/978-3-030-55169-8_8

1537 by the Flemish cartographer Gerardus Mercator, known for having invented a genial projection of the terrestrial globe based on the use of the compass, ideal for navigation, because it faithfully respected the angles between two lines at each point of the globe. And it was precisely in the field of cartography that Harriot achieved one of his best results, the brilliant solution to a problem that fascinated mathematicians and navigators of the time: calculation of the length of a route that always intersects the meridians at a constant angle.

It was thanks to his passion for maps and navigation that he moved to London, to start the most exciting period of his life. London during Elizabeth's reign was among the largest cities in Europe, the political, cultural and intellectual center of the entire English empire. The long line of officials and nobles joined actors, playwrights, poets, artists and jewelers, and, altogether, they animated the life of the Court. The bustling banks of the Thames constituted the main transport artery between the various districts of the city and the privileged communication channel between London and the rest of the world. Harriot immediately became part of that world of sailors, adventurers and pioneers animated by great dreams. He came into contact with Sir Walter Raleigh, Queen Elizabeth's great favorite, and apparently also her lover: a man of great culture and a strong temperament, brilliant and witty, always dressed in elegantly styled clothes that made his beautiful presence stand out. Halfway between a soldier and a poet, Sir Raleigh was, at the time, the most powerful man in England, corsair of the seas by direct order of the Crown, definitely one of the most emblematic and colorful figures of the Elizabethan era. He had proved his courage on the battlefields of France and Ireland and during ocean crossings and, although he was not noble, he became a key figure in the English court.

In the Service of Sir Raleigh

Harriot entered the service of Sir Walter Raleigh in 1584, moving to Durham House, where his protector was also staying, in a room located just under the roof; here, at night, he could make his beloved astronomical observations, while during the day, he continued his studies of mathematics and geography.

At the time, England was not yet a true marine power, especially compared to Portugal and Spain, both of whom had been sailing the oceans for centuries. British sailors were mostly familiar with offshore sailing, suitable for fishing or maritime trade operations on a local basis. In order to navigate the open sea, it was, however, necessary to know well how to measure latitude and longitude, to have updated nautical maps and precise compasses, tide almanacs and ephemeris of planets and various stars. It was for this reason that Sir Raleigh asked him to join him, because he knew that this young brilliant boy would be able to face all of that series of nautical problems with which he was concerned. And it was also for this reason that he wanted him in the second expedition to colonize the new lands of North America organized in 1585.

Once he arrived on the new continent, his curiosity shifted from mathematics to anthropology: not only did he learn the language of the indigenous peoples, he also assimilated its pleasant customs, not least of which was the delicious one of smoking tobacco in large cane pipes. Thus was born *The Brief and True Report of the New Found Land of Virginia:* for its detailed description of the flora and fauna, seashells and pearls of the seas, the various species of fruit and vegetables—including sugar, tobacco and pepper—and its dedicated study of the social customs of the local population, this book still remains a true jewel of ethnography and cultural anthropology. Harriot loved to enchant the indigenous peoples by showing them the marine compass, converging lenses, bursts with gunpowder, sundials, and so on. These were the tools that had changed the technology of Elizabethan England overnight and, watching them, the Native Americans believed that they were the work of the Gods and not of men, and this led them to think that the Gods must particularly love those white men. The reverence they had for Harriot's magic made it easier for them to talk about religion, and for the Europeans to convert them and keep them subjugated out of fear of the prodigious effects they showed them. To amaze the Native Americans, Harriot would sometimes line up a series of lenses in such a way as to concentrate the sunlight at a point further ahead and make a flame "rise" out of nowhere in the middle of the bush: the magic of the white man was used to bring the Sun to Earth... At other times, he had them look closely through his diverging lenses, and wow! Things now seemed now gigantic to them. Then, by moving a mysterious device, everything appeared upside down: trees, people, things, even animals had their heads down and their legs in the air. All of these tricks, and others facilitated with magnets and pieces of metal, contributed to the wonders and superstitions upon which the colonialism of those years was based.

The fleet, made up of a hundred men led by Sir Richard Grenville, sailed from England on April 9, 1585, and landed on the Virginia coast in late June of the same year. On the occasion of that trip, Harriot wrote a text entitled *Arcticon*, which, like his other books, was never published. It is a useful compendium of all of the orientation techniques that he studied at sea and then applied during navigation, such as determining latitude with measurements of solar and stellar amplitudes or measuring the variation of the Earth's magnetic field. On board of the ship the Tiger, crossing the Atlantic Ocean, he also had the opportunity to observe a solar eclipse, and the accuracy of his measurements allowed us to determine in the modern era exactly where the ship was on that day!

The Rectification of the Equiangular Spiral

Thomas Harriot first calculated—and in a brilliant way—the length of a curve that always intersects the meridians at a constant angle. This curve, Harriot established, gives rise to an equiangular spiral, that is, the curve found in nature in the elegant shapes of the Nautilus shell or in the coils of sunflowers. The problem, equivalent to that of calculating the length of a circular peeled orange peel, is easy to solve today, as we have the powerful means for infinitesimal analysis. At the time, however, very little was known about both differential and integral calculus, and the solution found by Harriot still leaves us amazed at the acumen that it shows: in fact, he obtained the rectification of the curve (i.e., the measure of its length) with purely geometric methods, noting that the triangles obtained by tracing rays starting from the origin of the curve, however small, can always be recomposed in the same, single large triangle, thus easily determining its length.

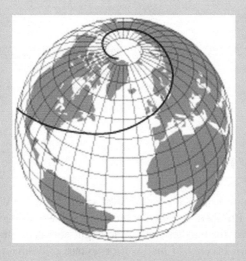

The construction of this straight line for a spiral is reproduced on the next page: this curve, Harriot noted, can be approximated by a polygon with sides s1, s2, s3,..., which, once connected to the origin p of the spiral, give place in the sequence of triangles T1, T2, T3,... These triangles T1, T2, T3, ... can be subsequently assembled to form the triangle ABT, whose area is evidently equal to that of the spiral. On the other hand, we have

BT + TA = s1 + s2 + s3 +··· = length of the spiral.

The approximation of the initial spiral can be improved more and more by progressively reducing the length of the sides s1, s2, s3,... Once the corresponding triangles have been recomposed—concluded Harriot—the result is invariably the same, i.e., the triangle ABT, for which both the area and the perimeter can be easily calculated, since it is an isosceles triangle with base a and a known base angle (in the example, 55°). Brilliant!

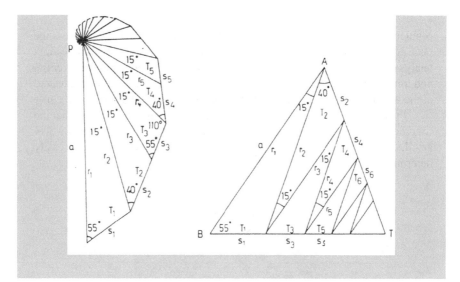

In the Tower of London

Upon his return home, however, a big surprise awaited him: Sir Walter Raleigh, the man whom Queen Elizabeth adored and for whom she was moved to tears when he spread his cloak over the puddles so that she might not get her feet wet, had fallen into disgrace, ruined because of his love for a court bridesmaid. "Hell knows no anger like that of a scorned woman," remembered the playwright William Congreve, and that of Queen Elizabeth was truly terrible: Sir Raleigh was indicted on suspicion of atheism and imprisoned in the Tower of London, while the investigation widened into a wildfire that threatened to engulf the whole circle of people around him.

For Harriot, this was enough to cause him to look for another mentor in a hurry: his choice fell on Henry Percy, Duke of Northumberland. However, never had a decision been more hasty and accompanied by risk: the Duke, in fact, only managed, just in time, to assure Harriot an annual income and the use of his residence in Syon as a home and workshop before he too was overwhelmed, as had been Sir Raleigh, by a judicial vortex that ended with him rotting in prison for over twenty years. This time, however, the cause was much more serious than that of the gallant Sir Walter; the accusation that fell on the Duke of Northumberland's head was, in fact, the participation in a plot to blow up London's Parliament with gunpowder. In the raid that followed the investigation, among the people arrested were Thomas Harriot and his friend Cristopher Marlowe, the cursed playwright, who—as did many other—had fallen from grace, from the privileged role of spy for the Crown to that of a miserable offender for such crimes as blasphemy and homosexuality. Harriot got away with a month in prison; Marlowe, instead, like one of the maudit characters of his own theatrical comedies, escaped the rigours of the law by dying, murdered in a sordid tavern in the backstreets of London.

Although his mettle had been much proved by his captivity and the continuous reversals of fate, Harriot returned to freedom and found the strength to continue his studies and experiments in the house of Syon. There, he would receive his mathematical friends Walter Warner and Thomas Hughes, with whom he formed a very close-knit trio: they were called "the three wise men of Northumberland." With his two colleagues, he discussed the discoveries that he was gradually making in various fields of science. His knowledge was increasingly articulate and profound, and his curiosity opened up new horizons of knowledge: besides being a master of the classic synthetic methods of geometry of Euclidus, Apollonius and Pappus, he proved also to be a very skilled expert in the field of algebra, studying the texts of Gerolamo Cardano, Rafael Bombelli and François Viète. And, indeed, some fundamental progress in modern algebra can be credited to Harriot, such as the introduction of the modern notation of algebraic equations and the ingenious solution of some equations of a higher order than the second. The pages of his notes were full of tables, formulas and multitudes of symbols, but they were very sparse with explanations, since Harriot wrote for himself and his friends, who somehow already knew the content of those notes. In this choice, of course, he was influenced by the historical and cultural context.

The scientific spirit of the time, characterized by a profound anxiety and a disconcerting instability, can be perfectly captured by following the steps of another singular character, Francis Bacon. Of all of those who hung out at the court, he was one of the most ambitious and cunning. He was born in London in the same year as Harriot, and the two probably knew each other personally, not least because they inhabited the same basic environment. Bacon's life was also marked by ups and downs: he undertook a very rapid career within the bureaucratic apparatus of the kingdom, which led him to assume the position of Lord Chancellor; however, as he had risen rapidly to the top, so did he descend just as quickly. He was accused of corruption and embezzlement, and was tried and sentenced to a very high fine and a period of detention in the Tower of London. The terms of his sentence were later mitigated, but his public career was over. Francis Bacon retired to private life, devoting all of his time to studying, focusing, in particular, on the scientific methods of empiricism. His fame as a philosopher of science was later affirmed with the publication of his masterpiece *Novum Organum Scientiarum*, which laid the foundations for the scientific method that would come to be most closely associated with Galileo Galilei. Bacon's story is full of legends and controversies about his alleged homosexuality, about the hypothesis that he was the illegitimate son of Queen Elizabeth and the Earl of Essex, about the idea that he was the true author of Shakespeare's works; he has been implicated as belonging to various secret societies and it has even been suggested that he staged his own death. However, despite his many contradictions and turbulent life, he managed to pave the way for the future of science, just as our Thomas Harriot was doing, almost simultaneously.

A Versatile Scientist

Harriot's interests were not limited to mathematics: like his contemporaries Tycho Brahe, Johannes Kepler and Galileo Galilei, with whom he was in correspondence, Harriot wanted to turn his gaze to the sky. Using the telescope—invented precisely in those years—he began producing the first lunar maps. He calculated the orbit and period of Jupiter's satellites, and became passionate about studying the phases of Venus. He also managed to determine the period of rotation of the Sun by observing the movement of sunspots, and he witnessed the passage of a famous comet, now known as Halley's Comet. He also became interested in optics, arriving at the formulation of the law of refraction in 1601, twenty years in advance of the "discovery" of what is known today as Snell's law, or, in chauvinistic France, as Descartes' law. On this subject, he had an interesting correspondence with Johannes Kepler, who was already residing in Prague at that time: in a letter of October 2, 1606, Kepler wrote to him that he had learned of his wonderful progress in the field of natural philosophy and optics, and he asked for his opinion on a series of themes, such as refraction, the halos of light that surrounded the sun and the rainbow. The fascination of the rainbow had not failed to attract Harriot's attention, and he had tried his best to come up with a mathematical explanation by setting up several optical experiments. His versatility also manifested in the field of mechanics, in which he correctly established—before Galilei's fundamental studies—that the trajectory of a projectile consists of a parabola and that the maximum range is obtained by directing the cannon at 45°. Finally, his work on the problem concerning the optimal storage of cannonballs in the holds of ships dates back to the time of his partnership with Sir Raleigh, work that would eventually be incorporated into what would later became known as the "Kepler conjecture" (see the chapter dedicated to him).

Today, Thomas Harriot is offhandedly remembered with a bronze plaque at the main entrance of the Bank of England, on the south side of Threadneedle Street in London. Originally placed in the basilica of Saint Christopher in which he was buried, the plaque managed to escape the destruction of the church in the great fire of London in 1633, and later found a new, albeit casual, location at the entrance of the banking institute of His Royal Highness that rose in its place. The joke of fate followed Harriot, not only during his life, but also after death. Harriot was thus forgotten, partly because his work was not published during years that might have been most useful for its dissemination. This, despite the fact that, in his will, addressed to his friends of the clan of the Magi of Northumberland, his desire was made very clear: he wished that all of his writings, collected in the residence of Syon, would be published in the form in which they were, and as quickly as possible. And that's friendship for you.

Further Readings

R. Arianrhod, *Thomas Harriot. A Life in Science* (Oxford University Press, Oxford, 2019)

C. Hoffman, *The Murder of the Man Who Was Shakespeare* (Grosset & Dunlap, New York, 1960)

S. Ronald, *The Pirate Queen: Queen Elizabeth I, Her Pirate Adventurers, and the Dawn of Empire* (Harper Collins, New York, 2007)

Ising. A Magnetic Modesty

When Nature Changes Mood

Phase transitions are among the most fascinating phenomena in Nature. The world would be terribly flat and boring if the different phases of matter did not exist and if we did not witness the continuous flow of their transformations. It is extraordinary how such metamorphoses occur in the mere changing of the temperature: the lower it drops, the more we enter a world full of surprises and amazing behavior. If we lived, for example, on the surface of the Sun, we would not know of the existence of many aspects of the natural world: in fact, at those temperatures, the only stable state of matter is the gaseous one, and we would have no way of observing the existence of either liquids or solids, much less would we be able to admire the beauty of snowflakes. In fact, it is only by lowering the temperature below the interval familiar to us that we begin to appreciate the multiple states of aggregation of matter, such as the fluidity of water or the wonderful regular shapes of crystals. But this is only the beginning: in fact, nature unfolds its full range of possibilities only by descending to ever lower temperature values. Around $-200\,°C$ (and for lower values), for example, many materials become superconductors, i.e., able to sustain an electric current indefinitely, even in the absence of an electric field, and to expel from its bulk every possible magnetic field. Liquid helium, at a temperature of $-260\,°C$, instead presents extraordinary phenomena of superfluidity, managing to spill spontaneously from its containers or exhibiting, if put in rotation, regular structures of vortexes. Further lowering the temperature, until we almost reach so-called absolute zero, we see in some gases the very peculiar phenomena of condensation, theoretically predicted at the beginning of the last century by Bose and Einstein and experimentally confirmed in 1995 by Eric Cornell, Carl Wieman and Wolfgang Ketterle (as a result of which they won the Nobel Prize in Physics in 2001). When such a Bose-Einstein condensation occurs, the vast majority of the particles of a gas have a tendency to go into the minimum energy state of the system, and therefore there is the possibility of observing, on a macroscopic scale, the quantum behaviors of matter.

© Springer Nature Switzerland AG 2020
G. Mussardo, *The ABC's of Science*,
https://doi.org/10.1007/978-3-030-55169-8_9

The common aspect of all of these phase transformations, and, let's face it, the most mysterious aspect at first sight, is the cooperative nature of the processes that take place at certain critical points of temperature, temperature values that are not universal, but vary according to the substance. How do millions and millions of water molecules know how to organize themselves in the regular crystalline form of ice when the temperature reaches $0°$? And how do 10^{40} atoms simultaneously align in one direction when a piece of iron, placed in a magnetic field, magnetizes by falling below $1300°$?

The Success of a Model

Analysis of the macroscopic behavior of a system, based on the laws that regulate the microscopic dynamics of atoms, boasts a long tradition in physics. In particular, it is the battlefield of one of its leading disciplines, statistical physics. Starting from the work of its main promoter, Ludwig Boltzmann, statistical physics has gradually provided an incredible number of theoretical surprises over the years, allowing us to reach a full understanding of many physical phenomena. In the study of the phase transitions that take place in ferromagnetic materials, a privileged role is played by the Ising model: it describes, in simple mathematical terms, the magnetic moments of individual atoms and their interactions. The popularity of this model among physicists (and beyond, given the growing interest of mathematicians, biologists, economists and computer scientists) is simply evident from the numbers: from 1969 to 2019, there were around 21,000 scientific articles that, in one way or another, referred to it, for problems concerning actual statistical mechanics, neural networks, the folding of proteins, the optimization of economic resources, and so on. And the number of citations is destined to increase, given that, on average, the Ising model annually appears in about 800 articles.

In view of this large number of quotations, it is therefore paradoxical that the extraordinary fame of this model is not accompanied by a similar reputation for the scientist to whom it owes its name. So who was Mr. Ising? He is certainly not as famous as Einstein or Schrödinger, but his life was no less interesting than theirs. Humble and modest, Ising remained in the shadows for many years, ultimately becoming famous almost by accident. In a certain sense, what follows is meant to be a tribute to the work of that myriad of scientists who, without seeking notoriety and not out of any desire to excel, carry out their work every day with dignity and seriousness, with the sole purpose of contributing to the progress of science. Their idea, their article—who knows?—could be the turning point of a new research field tomorrow.

The Story of a Model

Ernst Ising was born in Cologne on May 10, 1900, where the father was a merchant. The family, of Jewish origin, had subsequently moved to Bochum, Westphalia, a mining town, where everything revolved around coal and steel. The town had grown

over the years, but without any regulatory plan, so the mines were immersed in the quiet of the countryside. Little Ising had a clean face and a sweet look, with which he tried to capture all of the things that he saw and imagined. His hair was always in good order, neatly parted on the right, and he wore dignified clothes. In that little world, life was a constant repetition of reassuring monotony: the ecclesiastical rites of the Catholic families of his school friends, the colourful uniforms of the Prussian soldiers who marched in the streets, the noise of the mills, the yelling of the Jewish merchants, along with the pace of the carters and workers.

After graduating from high school, Ising had to go into the army for a few months. By that time, he was nineteen, however, knowing that there was very little to do in Bochum, he decided to move to Göttingen to begin his university studies in mathematics and physics. Göttingen had always been the intellectual heart of Germany. In the field of mathematics, it had a prestigious history enhanced by the great names of the nineteenth century, primarily Carl Gauss, who had also been director of the local astronomical observatory, and his successors Peter Lejeune Dirichlet and Bernhard Riemann. When Ising arrived in Göttingen, the director of the Department of Mathematics was David Hilbert, engaged at the time in the study of general relativity, whose fundamental equations he had discovered independently of Einstein. His pupil a few years earlier had been Hermann Weyl, who introduced the notion of gauge theory in 1918, the basis of every theory of fundamental interactions in nature, such as electromagnetism and quantum chromodynamics. Among other things, Weyl had tried to unify the electromagnetic field and the gravitational field in a single formulation, but, unfortunately, his attempt was not successful. Another resident of Göttingen was James Franck, who, along with Gustav Ludwig Hertz, had, in 1914, conducted an experiment of the absorption of light by an atom, admirably confirming Bohr's atomic model and leading the two scientists to be awarded the Nobel Prize in Physics in 1925.

Ernst Ising spent two enchanting years in Göttingen, benefiting enormously from the sparkling intellectual atmosphere. He became interested in problems of statistical mechanics and was advised to contact Wilhelm Lenz, the most experienced scientist in that field, who was, however, in Hamburg. For the benefit of his thesis, he moved there. Lenz had previously studied with Arnold Sommerfeld in Munich, where he had obtained a doctorate in 1911, and where he had subsequently remained as an assistant. At the outbreak of the First World War, he was sent to the French border as a telegraphist. After the war, he returned to Munich, remaining there until 1920, when he received an offer to become a professor at the University of Hamburg. Working with Somerfield, Lenz developed a keen interest in magnetism, in particular, by trying to understand why some substances below a certain temperature become magnetized and others do not. For Somerfield and Lenz, the answer was to be found in the new quantum theory, the principles of which were combined with the considerations made in the past by Maxwell, Boltzmann and Gibbs about the statistical mechanics of an enormous number of particles. Magnetism had also attracted the attention of Pierre Curie, who, at the beginning of the twentieth century, had formulated the problem in extremely modern terms, based on principles of symmetry and symmetry breaking.

When Ising visited Lenz, the latter explained his vision of the problem to him: some atoms in nature have a magnetic moment whose origin is quantum in nature;

this magnetic moment can be represented as a vector, or an arrow, which, in general, is free to orient itself in any direction; however, in the presence of other similar atoms, the magnetic moment of each of them tends to orient in the same direction in which the others are aligned. A spontaneous magnetization can, however, only occur below a certain temperature, called the Curie temperature of the material. One must then explain, in the simplest possible way, the reason for this spontaneous magnetization in iron magnetites and pyrites.

Discussing the general case, they immediately realized, was complicated. Using the usual tactics of physicists to simplify a problem by trying to maintain its nature anyway, after much discussion, Lenz and Ising came to the conclusion that it was already interesting enough to understand what was happening assuming that the magnetic moments could be aligned only along an axis, let's say, with the arrowhead up or down. In this drastic oversimplification of the system, the infinite possibilities for the rotation of an arrow in a three-dimensional space were sacrificed on the altar of the "goddess simplicity" and modeled in terms of only two states, as if they were the two faces of one currency. But, even so, the model remained generically intractable from a mathematical point of view, because the number of neighbours with whom a magnetic moment interacted could be very large, and this made life impossible.

There was one case, however, that could be solved exactly! It was the model in which the crystal was thought of as strictly one-dimensional and the magnetic moments aligned only up and down, as discussed in 1924 by Ising in his doctoral thesis. In itself, his conclusion was somehow disappointing, because he was able to show that, in the one-dimensional case, there never existed a Curie temperature such that, below it, a magnetization occurs. The reason he found was simple and could be explained qualitatively: imagine that all of the magnetic moments of the one-dimensional crystal are lined up, so that the system is initially magnetized. This state of order can easily be destroyed by a thermal fluctuation that induces all of the magnetic moments on the right (or left) of a given site to align downwards, creating a defect in the crystal. Ising had the opportunity to discuss his findings at length with Wolfgang Pauli, who, at the time, was teaching in Hamburg. Ultimately, his discoveries were published in 1925 in the journal "Zeitschrift für Physik," the prestigious scientific journal in which Einstein had published the works that gave him undying fame in 1905.

At the Mercy of History

After obtaining his doctorate, Ising went to live in Berlin and, during the years 1925 and 1926, he worked at the Patent Office of the Allgemeine Elektricitäts-Gesellschaft, following in the footsteps of Einstein. Dissatisfied with the job, however, he decided to embark on a teaching career, and for one year, he worked at a school in Salem, near Lake Constance. In 1928, he returned to the University of Berlin to study philosophy and pedagogy. The fatal attraction of the Germans for Goethe is well known: thanks to the legacy of his scientific studies, Ising's second—and last—article had the title

Goethe as a Physicist, and obviously addressed the great German poet's theory of colour theory.

After his marriage to Johanna Ehmer in 1930, Ising moved to Crossen an der Oder, where he became a teacher at the local high school. However, with Hitler's rise to power in 1933, the citizens of Hebrew origin were removed from public places, and so Ising lost his job, remaining unemployed for about a year. An era of flames and darkness was inaugurated, in which daily violence struck his neighbours, the shopkeepers on his street corner, and his closest friends, in an apocalyptic crescendo that opened gashes of terror throughout the society. The only salvation was to try to flee from that madness.

So, he went to Paris for a short time, working as a teacher at a school for immigrant children, but in 1934, he made the unfortunate decision to return to Germany, because he had found a new job as a teacher at a school opened by the Jewish community of Caputh, not far from Potsdam. In 1937, he became the dean. But the initial calm was immediately swept away by a dark cloud of violence that emanated a poisonous light, casting infernal shadows: on November 10, 1938, Ising was forced to watch helplessly as the school was destroyed by people incited by the authorities to follow the example of the general pogrom taking place against Jews throughout Germany. By the time darkness fell, only ash and rubble remained of the school.

On the Run to the United States

Those were terrible months, spent in anguish for the future and in terror of the present, which manifested itself in the stones thrown through the windows of houses, the hateful writing on walls, the yellow star affixed to many now-closed shops of the centre. And what Ising feared most soon came true: in January 1939, he was taken into custody by the Gestapo and interrogated for hours in dark barracks on the outskirts of Berlin.

Fortunately, after a few days, he was miraculously released, and his only thought was to escape from the abyss into which Germany had fallen. He no longer recognized his native land: the streets had been invaded by dark-shirted troops; in the evening, the locals were full of fanatics who praised the Führer; hatred and violence had penetrated people's skin, and had spread all the way to their hearts. Ising and his wife packed everything they could in two suitcases and fled to Luxembourg, with the intention of emigrating to the United States. But their asylum application was rejected, because of the limits imposed by the United States on migration flows. So, they decided to stay put, waiting for the first possible moment to obtain a visa, but the following year, on the day of his fortieth birthday, the Germans invaded Luxembourg and, with the consequent closure of all consular offices, every possibility of expatriation vanished. The fateful game taking place in Europe was once again playing havoc with the life prospects of many people.

Captured once again by the Nazis, Ernst Ising managed to escape the heinous end of many other Jews for the sole reason that he was married to an Aryan woman. In exchange for his life, he was induced to forced labour and assigned to the dismantling of the fortifications of the Maginot line. Surviving multiple horrors purely by chance, two years after the end of the war, Ising and his wife finally managed to board a cargo ship bound for the United States.

There, he initially taught at the State Teacher's College in Minot, and then moved on to Bradley University, where, from 1948 until 1976, he held the chair of Physics. He acquired American citizenship in 1953, and in 1971, Ising was awarded the Outstanding Teacher of America award. Having avoided death in Nazi Germany, Ising succumbed to his mortality the day after his 98th birthday, May 11, 1998, at his home in Peoria, Illinois.

A Lesson in Great Honesty

Like many other German physicists, Ernst Ising's life and career were deeply marked by the Nazi dictatorship and the dramatic episodes of the World War. After his graduation thesis, he never returned to active research, thus remaining unaware of the scientific developments of the time for many years. A different fate, however, awaited his article. Cited for the first time in a 1928 work by Werner Heisenberg, its fame was consolidated thanks to a famous 1936 study by Rudolf Peiers, entitled *On the Ising model for ferromagnets*. From that moment onwards, scientific literature has seen the uninterrupted proliferation of a large number of publications about this model, which saw an especially large increase after the ingenious solution of the two-dimensional case proposed by Lars Onsager in 1944.

To give an idea of the importance of this result, it is sufficient to mention the correspondence between Wolfgang Pauli and Hendrik Casimir, which took place immediately after the Second World War. To a letter in which Casimir expressed bitterness at being unaware of what had happened within theoretical physics during the war, Pauli's lapidary answer followed: "Nothing important happened, except the exact solution of the two-dimensional Ising model by Lars Onsager." He only omitted mentioning the fission of the atom, the control of nuclear forces, the atomic bomb and even the Nobel Prize received by Pauli himself in 1945!

In closing, it is also worth adding that it was only in 1949 that Ising discovered—with some surprise—the notoriety of his work and the fame that his name had acquired within the vast scientific community. However, his modesty led him to specify, in a letter to his biographer Sigismund Kobe, "I would like to emphasize that the model should correctly be called the Lenz-Ising model. He was my supervisor. Prof. Wilhelm Lenz had the idea for the study and proposed analyzing the mathematical content as my thesis work." Hat's off, Prof. Ising, indeed, for this last great lesson in fairness and honesty.

Further Reading

S. Brush, History of the Lenz-Ising Model. Rev. Modern Phys. **39**(4), 883–893 (1967)

Joliot-Curie. The Dream of the Alchemist

There were only a few days before Christmas, and the streets of Paris were full of sleet: from the river, a freezing wind was blowing, and everyone seemed to hurry as if looking for shelter. Some carriages stood in front of Notre Dame, the coach drivers intent on calming the horses that were moving about restlessly. Along the Seine, a few barges full of coal passed by, heading for the small berth immediately beyond Pont Neuf. Frédéric Joliot hurried through Pont de l'Archevêché, turning onto one of the streets of the 5th arrondissement that led to the Institut du radium.

He was a tall and slender, with a thick mass of black hair combed back, dark eyes full of energy, and the athletic attitude of a twenty-four-year-old man accustomed to being outdoors, running, playing sports. That December 16, 1924, he had an appointment with Marie Curie at nine 'o clock in the morning: there seemed to be the possibility of an assistant position—as Paul Langevin, his teacher and great friend, had told him—and he would not have missed the opportunity for all of the gold in the world. For years, he had had a photograph of Pierre and Marie Curie in his bedroom near his bed—his idols since he was a boy! And now, perhaps, he would have the opportunity to work with the great scientist who had revealed the mysterious world of radioactivity, made of substances that shone in the dark and warmed the fingers to the touch.

He arrived at the Institut du radium a little out of breath, knocking on the door and trying to mask his agitation by pretending to play with his hat, which he passed from one hand to the other. Marie Curie was waiting for him in a small study, which contained a table with sheets scattered all across it, a desk covered in books of chemistry, a bottle of ink, several pens and a Geiger counter placed sideways, and shelves that ran along the walls housing various glass ampoules. He had been told that she looked older than her fifty-seven years old, yet Joliot was surprised to see that she was so small, with completely gray hair, her fingers worn by the fire: her eyes seemed very sweet, but her bearing was that of a proud and severe-looking woman. She definitely put him in awe. After a few marginal pleasantries, she immediately began to discuss his school curriculum, and was somewhat surprised to learn that

© Springer Nature Switzerland AG 2020
G. Mussardo, *The ABC's of Science*,
https://doi.org/10.1007/978-3-030-55169-8_10

Joliot did not yet have a degree. She reminded him that a degree was necessary for anyone who wanted to pursue a career in the world of science.

"You are right," said Joliot, "I'm going to do it soon."

"Would you be willing to start working tomorrow?" Marie Curie suddenly asked, looking him in the eye.

"Do you really mean it? Unfortunately, I still have three weeks of military service to finish. I don't know if I will be able to …" he replied, with a note of strong regret in his voice.

"Don't worry, I'll personally write to your battalion commander to get his permission for you to come here starting tomorrow."

So it was. The next day, Frédéric Joliot was already bent over the laboratory workbench. Her faith in him had not come out of nowhere: Marie Curie was always very attentive to what Paul Langevin said to her, and if he said that the young man had a bright future ahead of him, well, she had to believe him. Her work took up so much of her time; perhaps now, having engaged some help, she might have a moment or two to sit back and reflect upon the many memories that came back to her: the old gray house in Warsaw where she was lived as a child, her father's face, the streets she used to travel in her youth, the first impressions she had of Paris, as a great city for theater …

Glory and Tragedy

Marie Skłodowska had first arrived in Paris in 1891, after spending her adolescence in Warsaw, during a period in which Poland was under Russian domination. As a child, she had experienced the premature deaths of both her mother and her sister, two extremely severe traumas that only enhanced her already existing sense of melancholy and soured her on religion for the rest of her life. After school, she had spent some years as a housekeeper with several families, earning a reasonable wage and using her free time to learn French on her own. But when she was twenty-four years old, she felt that the time had come to take her life into her own hands. Collecting her few belongings and the books from which she was never separated, she left for the longest journey that she had ever made. It took her three days to reach Paris by train, where she wished to enroll in the Sorbonne, the temple of physics and chemistry, the two subjects that had always fascinated her most.

Shy and somewhat bewildered, but with exceptional determination, she dove right into the studio, the only woman in that large academic ocean of men. She spent whole afternoons sitting at the long rectangular table in the Library of Sainte Geneviève, her head in her hands, immersed in the reading of manuals, so detailed, so interesting, as if in the grip of a fever. She graduated after only two years, magna cum laude. A single degree was obviously not sufficient, and in addition to physics, she also wanted to obtain another one in mathematics. It took her only a year to achieve this as well.

In the fall of 1894, she returned briefly to Poland to see her elderly father, her friends and her dear sisters, but in her heart, she knew that the centre of gravity of her life was now located in Paris. Among other reasons for this was the fact that she had met Pierre Curie a few months earlier, a chemist, a very original man, eight years older than her. His face was illuminated by an incomparable gaze, as deep and serene as his soul. Pierre had totally dedicated his life to research, and in his laboratory, he had discovered the most strange phenomena: the appearance of an electric charge on the surface of crystals if they were compressed or dilated, the magnetic domains in materials such as iron or nickel, which appeared below a certain temperature, and the role played by symmetry in the laws of nature. He had spent many years studying the subject, having fun building his own laboratory tools. When they met, their conversation was dominated exclusively by science: Marie fervently exhibited some experiments that she wished to carry out, Pierre described to her the discoveries that he had made in the field of crystallography and electricity. "We started a conversation that soon became friendly," she wrote years later.

Pierre was fascinated by this Polish girl, as devoted to science as he was. He proposed that she come to work at the laboratory that he directed at the École supérieure de physique et de chimie industrielles: the rules of the French schools did not allow women to be paid, but Pierre was happy to take charge of the payment personally. The two soon became inseparable. They married in 1895 in a civil ceremony, deciding not to exchange wedding rings and to use the money from their gifts to buy two new bicycles: their honeymoon was a long trip through the Brittany hills, visiting the villages along the Normandy coast. In a photo taken at the time of their departure, she wore a straw hat, a wide blouse, knee-length pants and ankle boots, while he wore a jacket and sports trousers: both had a serious but radiant look, the very image of happiness.

Since then, Marie had taken her husband's surname, but she was always very attentive to her independence and to her realization as a woman and a scientist. Their first daughter, Irene, was born in 1897. Immediately afterwards, Marie was convinced that the natural outlet for her studies would be to pursue a doctorate: fortunately, Pierre's father, following the death of his wife, had decided to move to their home on rue de la Glacière and to take care of the child. Marie was interested in an article written the previous year by Henri Becquerel in which he reported the strange phenomenon of a spontaneous emission of radiation by a sample of uranium salts, whose traces remained on a photographic plate. Becquerel also noted that the uranium sample easily ionized the air, as could be seen from the movement of the lamellae of an electroscope placed nearby. These were the first phenomena of what is now known as radioactivity, a name coined by Marie Curie herself. Becquerel's discovery greatly fascinated the Curies: what was the nature of that radiation? Where did all of that energy come from? Undoubtedly, it seemed like the right subject for a research thesis and promised to be a great adventure, in a completely unknown field of knowledge. Going deeper and deeper into the analysis of that radiation, Pierre and Marie Curie realized that it was a phenomenon that was widely independent of the temperature conditions and chemical reactions to which the uranium salts were subjected: in simple terms, the radiation emitted by uranium seemed to depend only on the properties of its atomic nucleus. Marie also took another step forward: was it possible that uranium was the only element with these properties? Were there other elements capable of radiating? It was certainly appropriate to investigate further.

From the Czechoslovakian deposit at Jáchymov, Marie and Pierre Curie purchased a few tons of pitchblende, a mineral from which uranium was typically extracted, and they immediately noted that some samples were much more radioactive than those consisting of uranium alone. At the beginning, they thought that it was a mistake, and they repeated the experiments once, twice, ten times, always coming to the same conclusions. Maybe they were in the presence of a new chemical, one of those boxes that remained empty on Mendeleev's table! The improvised laboratory on rue Lhomond thus became the theater of one of the most beautiful scientific enterprises of all time: "We walked up and down, always talking about our work, present and future. When we were cold, a cup of tea drunk near the stove cheered us up. The research absorbed us continuously, so much so that it seemed we were living in a dream." After a two-year period of tireless work, passed among toxic fumes, crucibles, test tubes, filters, flasks and Bunsen burners, the Curies finally managed to isolate some bismuth compounds from the pitchblende with radioactivity 400 times higher than that of uranium: "We believe that the substance we extracted from pitchblende contains a metal that has not yet been reported. If the existence of this metal is confirmed, we propose to call it Polonium." But the mere presence of polonium and uranium did not yet seem to account for the enormous radioactivity that they had found in the samples. There was only one possibility: the existence of another element! In the laboratory notebook, dated March 28, 1902, Marie Curie noted: "Ra = 225.03. Atomic weight of Radium." At night, the bluish light that came from the tubes with the salts of radium brightened the laboratory room, creating an almost enchanted atmosphere.

In 1903, the Curies received the Nobel Prize for Physics, together with Henri Becquerel, for their studies on radioactivity. In the blink of an eye, husband and wife became very famous. Pierre Curie was appointed professor at the Sorbonne, but everyone seemed most intrigued by Marie, a female scientist! And along with the fame of the Curies, radio madness had also broken out throughout the world, perhaps because it shone in the night, or because it was imagined to cure diseases, or simply because it was huge news: rightly or wrongly, it began to appear on the labels of many products, whether they were shoe polishes, toothpastes or detergents. Notoriety did not detract from the Curies' strong ethical sense, and they decided not to patent their process of isolation of radium in order to allow the scientific community to research the field of radioactivity freely. In 1904, their second daughter, Eve, was born, essentially crowning a series of exceptional years.

However, tragedy was just around the corner: on April 19, 1906, travelling along rue Dauphine after leaving a meeting of the Academy of Sciences, where he had been more enthusiastic and talkative than usual, Pierre Curie slipped and fell under a horse-drawn carriage, which ran him over, killing him instantly. For Marie, it was an excruciating pain, a violent punch in the stomach; in the months that followed, she devolved into obstinate silence, refused to eat, and slept little and rarely. The only thing that eventually convinced her to move forward was her daughters, each more adorable than the other. The Sorbonne offered Marie the post of professor left empty by Pierre: she was the first woman to have a teaching post at the University of Paris. On November 5, 1906, in front of a classroom full of students, she gave her first lesson, moved to tears, picking up exactly where Pierre had left off.

After the summer of 1911, people began to notice a sudden change in Marie: her face became much more serene, her eyes shone, and she had abandoned the black dress—which she had begun to wear on the day that Pierre died—in favour of a light dress and with a large beautiful flower at shoulder height. It was clear that there was a new person in her life, a man with whom, like Pierre, she shared the ideals of life and social and political beliefs. The scandal that this caused erupted with unprecedented violence in November 1911 and lasted throughout 1912: Marie was accused of having an adulterous relationship with Paul Langevin, who had been a student of her husband and who was part of the circle of the family's friends, a married man, albeit unhappily, with four children. Marie and Paul had exchanged letters full of tenderness and passion, letters that had ended up in the hands of Langevin's wife and, directly from her, the pages of the newspapers, especially those of a conservative bent, happy to give a hard time to a woman, especially a foreigner. In the previous months, those same newspapers had orchestrated a brutal and angry defamatory campaign against Marie Curie, on the occasion of her possible appointment to the Academy of Sciences, accusing her of everything: of being a careerist, of having exploited the fame of her late husband, of being Jewish (she, who was actually from a Catholic family), going so far as to invite her to leave France and return to Poland. She could have been the first woman elected to the Academy, but the vote of the members unfortunately mortified her expectations. Since then, the climate had remained very tense, so much so that even the news of her second Nobel win in 1912, this time for Chemistry, was relegated to a few lines on the inside pages of the papers. The scandal even reached the ears of the Nobel committee: Marie received a letter from Stockholm in which, between the lines, it was said that perhaps it would be better for her not to collect the prize. She firmly replied that she believed the prize had been awarded to her for her scientific work and that her private life was her business alone: she then presented herself at the ceremony in the company of her sister Bronia and her daughter Irene, then fourteen years old. It was an exciting experience for them, one that Irene remembered for the rest of her life, even though it would not be her last visit to Stockholm, at the Court of the King of Sweden. Marie's love affair with Paul Langevin did not ultimately work out, but the two always remained on excellent terms, as proven by the consideration with which Marie had welcomed Frédéric Joliot after Langevin had recommended him to her.

A Lucky Encounter

Frédéric Joliot was the sixth and last child of a modest family that nonetheless had solid roots in French political affairs: his father, Henri Joliot, had participated in the adventure of the Paris Commune, and like many other workers, had fought to the end for the revolutionary ideals that had animated the streets of Paris throughout 1871. Following the defeat, he had been sent into exile, and was only able to return to France in 1880, thanks to a general amnesty. At his death, the family had faced a serious financial crisis, which, among other things, had precluded Frédéric from attending the best schools. But that was perhaps to his benefit.

After high school, in fact, he enrolled at the *École supérieure de physique et de chimie industrielles*, the same school where Pierre Curie had worked years before and whose director was, at the time, Paul Langevin, who immediately noticed the young man's extraordinary qualities, in particular, his talent in the laboratory and the great dexterity that he showed with physics instruments and on chemistry benches. Langevin got into the habit of inviting him to his home on a regular basis: they talked for hours, about science, philosophy, politics; there was no topic on which they didn't have their say. A great friendship was established between the two, along with huge esteem and a mutual respect. As a result, Frédéric Joliot, from then on, turned all of his attention towards science and socialism, his greatest life choices. Langevin had advised him to study physics in particular, believing that this was his path; he was convinced that he had all of the qualities of a great scientist. And now, in Madame Curie's laboratory, he finally had the chance to prove it.

He was assigned a task that was far from easy, that of studying the chemical properties of polonium, a topic about which he knew almost nothing. This research was suggested by Irene Curie, twenty-seven at the time, who had been serving for several months as a laboratory assistant in rue Lhomond, a situation that also allowed her to work on her doctoral thesis. Irene's close relationship with her mother was all-encompassing, and had been cemented even more during the years of the First World War, when they had set up an ambulance service together equipped with machines that could take immediate x-rays of the soldiers wounded at the front, an experience that caused Irene to reflect thusly: "My mother never doubted me more than she doubted herself … I look a lot like my father, and perhaps this is the reason why we understand each other so much." Not at all intimidated by the difficulty of the task assigned to him, Frédéric threw himself headlong into that undertaking and, in a short time, he had discovered a whole series of new procedures for characterizing the chemical reactions in which polonium took part. He recognized, for example, that this element behaved like bismuth or tellurium: it was hardly soluble in an alkaline medium, but extremely reactive with acids.

Moving on with the experiments, he took note—with great satisfaction—that he had surpassed all of the expectations of the "Royal Princess," as Irene Curie was called in the laboratory. He noticed that she was now looking at him with other eyes, which he took as a sign that he had won her esteem, not an easy feat: in addition to her talent and incredible intelligence, which everyone acknowledged, Irene had the reputation of being an extremely detached, imperturbable person, cold, brutally honest, totally dedicated to work and light years away from any other interest. The Royal Princess simultaneously instilled respect and awe in all of her collaborators. Years later, Frédéric remembered that the first day he arrived at the laboratory, Irene had looked at him with a certain distance, perhaps even a bit of annoyance, as if it bored her to have to teach the newcomer all of those things that she had learned after so many years spent bending over the workbench. At the time, Irene was already part of that very small group of scientists with a very solid background in radiochemistry: the topic of her thesis, which she conducted under the supervision of Paul Langevin, focused on the study of alpha (α) particles emitted by polonium during radioactive decay, after which it turned into lead.

The thesis was discussed on March 30, 1925, and, a few weeks before the defense, Irene had given it to Frédéric to read, accompanying the request with a comment: she hoped that his reading was less boring than his writing. That day, at the Sorbonne, about a thousand people came to hear her dissertation, and many international news-papers, including the New York Times, covered the event: that the daughter of a great scientist, a two-time Nobel laureate, would also follow the path of science was more than enough raw meat for public opinion, and the opinion was expressed more than once, on that occasion, that she would be the next woman to win the Nobel Prize. Irene Curie showed up in the Sorbonne's lecture hall wearing a simple black dress, similar to what her mother had worn almost twenty years earlier, and, in a firm tone, she spent about an hour describing how she had learned to study the nuclei of atoms by bombarding them with the extremely energetic α particles emitted by polonium. She finished the presentation by dedicating her thesis "to Madame Curie, by her daughter and her disciple." As her sister Eve later wrote: "Irene showed up at the Sorbonne calmly, ended her presentation with the confident air of someone who knows that she has passed the exam and then waited without great emotion for the result of the commission," which was obviously very flattering. The reception was held in the garden in front of the Institut du Radium, and all of the people from the laboratory took part, including Frédéric Joliot, who later admitted, "I had discov-ered in that girl, whom many considered to be made of ice, an extraordinary person, very sensitive and poetic, which in many ways seemed like a living replica of what Pierre Curie had been. I had heard about him from many of my teachers and I was surprised to find in his daughter the same purity of mind, the same good disposition, his same humility." The two married on October 9, 1926, with both adopting the double surname 'Joliot-Curie.'

The Transmutation of Things

Frédéric and Irene, like Pierre and Marie years earlier, formed a formidable duo, both in their private and professional lives. They had two admirably complementary personalities: if Irene was meditative, Frédéric was rather more impulsive; if she liked to deal with the chemical aspects of the experiments, he liked the physical ones; if she was reserved and of few words, he loved to be in the midst of company and to fascinate everyone with his small talk. Nonetheless, as Irene wrote to a friend of hers: "We have many opinions in common on essential issues." Both shared, for example, a political passion for socialism and the working class; although they were pacifists and animated by a great humanitarian spirit, they loved France and were devoted patriots. For several years, they worked separately, but in 1931, they decided to tackle a truly intriguing question together from the field of nuclear physics, a sector in which competition was very strong due to the presence of a couple of very aggressive experimental groups, such as Ernest Rutherford's in Canada and Lise Meitner's in Berlin. The beginning of their adventure, it must be said, was not the happiest.

The question that intrigued them concerned the observation previously made by two German physicists, who had found that a very strong radiation was emitted by beryllium—a substance that is not radioactive in itself—when it was irradiated with extreme α particles of energy. As it was not deflected by magnetic fields, the German physicists concluded that this radiation was certainly not due to electrically charged particles, yet its energy was such that, directed on a paraffin layer, it gave rise to the escape of hydrogen atoms from the material at high speed. The Joliot-Curies repeated the experiment, taking advantage of their access to a source of α particles thanks to a sample of very pure polonium that they had synthesized, but they erroneously came to the conclusion that the radiation in question was made up of highly energetic gamma rays (γ).

Rutherford, with the sense of a great explorer of that atomic and nuclear world, had instead felt that there was another, far more revolutionary explanation, and therefore suggested to one of his students, James Chadwick, that he investigate further. Working madly night and day, Chadwick finally managed to solve the riddle: the radiation observed by the Joliot-Curies was due to a new particle, the neutron, with a mass approximately equal to that of the proton, but without an electric charge! For this fundamental discovery, which finally brought some order to nuclear physics, Chadwick was awarded the Nobel Prize in Physics in 1935.

However, it was not the only mistake that the Joliot-Curies made. In other experiments, they had noticed the presence of a particle with the same mass as the electron, but with a positive electric charge: the traces of the fog chamber, in fact, showed a spiral that was in all respects similar to that of electrons, only rotated in the opposite direction. Although Dirac's theory on the positron—the positively charged antiparticle of the electron—had been known since 1928, the two failed to interpret their data correctly: it was up to the American physicist Carl David Anderson to announce, shortly thereafter, the discovery of this new particle, found by analyzing the photographic plates with which he was studying cosmic rays at the time. Carl Anderson was awarded the Nobel Prize in Physics in 1936.

As the saying goes, however, third time's the charm. Bombing aluminum with α particles emitted from their polonium source, they also observed radioactive phosphorus accompanied by a neutron among the results of the reaction. When they communicated their discovery to the 1933 Solvay Congress, the reaction was not what they had hoped for: they were, in fact, openly criticized by Lise Meitner, who was the only woman, besides Irene, present at that conference, a woman whom Einstein had referred to as "the German Marie Curie." In front of the participants of the Congress, the elite of world of physics, which included, among others, Fermi, Rutherford, Schrödinger, Dirac, Heisenberg, Einstein and Pauli, Lise Meitner made all of her reservations about the interpretation of their data explicit, saying that, in experiments that she had carried out in the past, she had never noticed the presence of neutrons. However, Niels Bohr spoke up in defense of the Joliot-Curies, pointing out the importance of the question and the existing difficulties for a satisfactory solution, thus urging the two to simply repeat the experiment. "We'll talk about it again," were his final words.

Following Bohr's suggestion, they returned to repeat the experiment with utter perseverance, noting this time that, together with the neutron, there was also a positron, their two black beasts … But this time, the clue that they needed was in front of them: what they were observing before their eyes was, in fact, the result of two nuclear reactions in sequence. In the first reaction, as long as the aluminum was bombarded by α particles, it emitted a neutron, turning into radioactive phosphorus. In the second reaction, in the nucleus of the radioactive phosphorus, one of its protons turned into a neutron, emitting a positron, and thus transforming itself into a silicon nucleus. The net result of these two reactions was therefore the transmutation of two stable chemical elements: aluminum had become silicon!

"We had fulfilled the ancient dream of alchemists," wrote Frédéric Joliot-Curie years later, "that of transmuting matter." They showed the results of their experiment to both Paul Langevin and Marie Curie: the latter, her fingers completely damaged by radiation, did not hesitate to take the tube used by the two for their experiment and bring it close to a Geiger counter, which began to crackle with a thousand ticks. "That was, without a doubt, the last great satisfaction of her life," said Frédéric.

This time, the two were right, and were awarded the Nobel Prize in Chemistry in 1935 as a result. Unfortunately, Marie Curie had passed away the year before and was unable to rejoice in that last Nobel Prize, which had further increased the imperishable fame of her family.

Further Readings

E. Curie, *Madame Curie* (Doubleday, Doran & Company, 1937)
A. Pais, *Inward Bound* (Oxford University Press, Oxford-New York, 1986)

Kepler. Cannonballs and Honeycomb Cells

Johannes Kepler is one of the most atypical figures in the history of science. In fact, there is no other character who better embodies the passage from Renaissance magic to the dawn of modern science. Passionate about astrology and mysticism, prisoner of the religious and baroque culture of the time, he nevertheless managed to turn his gaze to the heavens, and what he ultimately saw paved the way for modern developments in science. The depositary of Tycho Brahe's extraordinarily accurate measurements of the position of the stars and the movement of the planets, Kepler discovered the elliptical orbits and a cosmos governed not by Aristotelian perfection, but by the laws of physics and geometry. The Copernican vision paradoxically found confirmation in the magical atmosphere of Rudolph II's Prague, which teemed with astrologers, fortune tellers and alchemists.

Before we deal fully with the famous geometric conjecture linked to Kepler;s name, it is worth getting to know this great scientist more closely. As his biographer Max Caspar wrote: "The nobility of his character brought him many friends, the vicissitudes of life attracted him sympathy, and the secret of his union with nature attracted the interest of those who wanted something deeper and different than what science could offer. In their hearts, they had devotion and love for this exceptional man. Because, once you entered the magical sphere that surrounded it, there was no longer any escape." His life was troubled by serious family problems, illnesses and major financial problems. However, in spite of all these difficulties, Kepler managed to climb the highest scientific peaks, from which he could then look at the harmony of the universe. When Newton wrote, "If I have seen further, it is because I have stood on the shoulders of giants," one of the giants to whom he was referring was indeed Kepler.

© Springer Nature Switzerland AG 2020
G. Mussardo, *The ABC's of Science*,
https://doi.org/10.1007/978-3-030-55169-8_11

The Astrologer Astronomer

Johannes Kepler was born on December 27, 1571, in the town of Weil der Stadt in Germany. His father was a soldier of fortune who used to come and go from home according to the wars in which he was involved, habitually exhibiting a violent and unpleasant demeanor when at home. Kepler was therefore essentially raised by his mother Katharina, a woman with a sharp tongue and a witty mind, who mostly dedicated her life to distilling herbs and medicinal potions in a corner of the house, a hobby that, years later, would become the source of serious problems. The family moved to Würtemberg and, in 1588, the father walked out for good, never to be seen again: it was said that he may have died in a battle or possibly on the way home from one, nobody ever knew for certain. As a child, Kepler was sickly: he barely managed to survive an infantile illness, which, however, damaged his eyes, so that he was forever denied the pleasure of making astronomical observations. After attending a monastic school in Maulbrom, at seventeen years of age, he enrolled at the university of Tübigen, ready to embrace an ecclesiastical career. But fortunately for him, and for all of us, he was influenced by Michael Maestlin, a professor of mathematics at that university and an ardent supporter of the Copernican vision. Maestlin not only fully understood the value of his pupil, but, at the right time, he also helped him to get a chair of mathematics at the University of Graz, in the Styria region. It took Kepler about a month to move from Bavaria to that southern region of Austria: it was a place that seemed remote to him and was overwhelmed by serious religious tensions between Catholics and Protestants. In Graz, he found a wife, but the marriage very soon turned out to be an extremely unhappy union that was made even unhappier because of the bad economic conditions that he was facing, having to provide for the welfare of his wife and three children, who had arrived in the meantime. He rounded out his revenue by making astrological predictions, an activity that accompanied him for many years and about which he always had mixed feelings: if, from a rational point of view, he recognized the superstition of the thing, from the mystical point of view, he did not hesitate to believe that the alignment of the planets or the appearance of a comet could have great consequences for human life.

Harmonices Mundi

In Graz, as concerned science, Kepler spent his years very happily, immersed in deep speculations on the solar system that straddled both the noble art of geometry and Renaissance hermeticism. At the time, only six planets were known to exist in the solar system: starting closest to the Sun, Mercury, Venus, Earth, Mars, Jupiter and Saturn. Furthermore, their average distances from the Sun were also known with some precision. Between six bodies, there are obviously five spaces that separate them, five also being the number of regular Platonic solids! On July 20, 1595, Kepler thought that he had found the key to the skein and explained his discovery with the

following words: "The sphere of the Earth is the measure of all things. We circumscribe a dodecahedron around it. The sphere that will inscribe this solid will be that of Mars. We circumscribe a tetrahedron to it. The sphere that will inscribe this solid will be that of Jupiter's orbit. We now circumscribe a cube around Jupiter. The sphere that will inscribe this solid will be that of Saturn. Now, let's inscribe an icosahedron to the Earth sphere. The sphere that will inscribe this solid will be that of Venus. Finally, we inscribe an octahedron within the sphere of Venus' orbit: the circle inscribed inside this octahedron will be that of Mercury's orbit." The result of this construction is the famous image known as *Harmonices Mundi*, which Kepler included in the book *Mysterium Cosmographicum* in which he exposed the following theory: the distance between these boxed Platonic solids seemed to correspond magically to the various distances of the planets from the Sun!

The Dane with a Golden Nose

Tycho Brahe was a singular Danish nobleman, with a proud look and a golden nose, literally, as he had lost the real one in a duel that had followed from an animated discussion with a fellow, not about a woman, but about which of the two had more mathematical talent! He was the grandson of Vice-Admiral Jørgen Brahe, famous for having saved the life of King Frederick II of Denmark, who had fallen into a Copenhagen canal. This episode in the family's history later smoothed the way for favours by the court and the support of the king. Since he was a child, he had cultivated a passion for astronomy and, studying Ptolemy's *Almagest* and Copernicus' *De revolutionibus*, he realized that the astronomical hypotheses about which everyone was arguing at the time were based on antiquated astronomical tables that gave unreliable information on the orbits of the planets. To arrive at some satisfactory conclusion, he understood that it was therefore necessary to rely on instruments that were as precise as possible. In 1575, upon returning from a long trip to Venice, the King of Denmark commissioned him to build an astronomical observatory on the island of Hven, in the middle of the North Sea. Here, Brahe built an observatory castle, which he dedicated to the muse of astronomy, Urania, calling it "Uranienborg." That island, perpetually beaten by the freezing winds of the sea that separates Denmark from Sweden, quickly became the site of an advanced research centre, one of the first that modern science remembers. Here, Brahe proved to be an extraordinary observer, making increasingly precise measurements of the stellar coordinates and the motion of the planets and comets. He built new instruments and, thanks to a huge and very accurate sextant, the Danish astronomer managed to measure the positions of different celestial stars with unparalleled precision for the time. A decidedly surprising result, if you consider that Brahe's measurements were all made with the naked eye. In the golden age of Uranienborg, the island castle also housed a printing press for publishing the results related to the various astronomical observations: of these, the most important one was undoubtedly the observation made by Brahe on November 11, 1572, concerning the explosion of a large star in the constellation of

Cassiopeia. Nowadays known as SuperNova 1572, or even Tycho Nova, the event signaled the violent death of a star. What was left of that celestial body remained visible to the naked eye until March 1574, allowing Brahe to note that the parallax of this bright object had never changed during the various observations, thus leading him to the conclusion that such a bright star must be very distant from Earth. Supernova or comets, the sky was therefore less perfect than what Aristotle had thought...

Brahe nevertheless embraced neither the Ptolemaic model nor the Copernican theory, but rather proposed his own, singular model of the Solar System, according to which all of the planets rotate around the Sun while the Sun and the Moon rotate around the Earth. In 1597, due to disagreements and cuts in his funding made by the new king of Denmark, Christian IV, Brahe moved first to Wandsbech, Germany, and then to Prague, where he was summoned by the emperor Rudolf II.

Box: The Wisdom of Bees

Kepler's interest in packing the spheres in a three-dimensional space stemmed from his curiosity about the shape of snowflakes. These differ from one another in many details, but they all share a general property: that associated with their hexagonal symmetry, that is, by turning a snowflake 60° (or one sixth of a complete turn), one finds that it looks exactly the same.

To understand this curious property, Kepler resorted to the geometry of which he was so very fond. He then made the hypothesis that solids in nature tend to completely fill the space, and that a gedankenexperiment to understand the shape that they take would consist of starting from a configuration of spheres and imagining that these can expand in order to fill all of the interstices. Thanks to this reasoning, we can indeed understand the shape of the grains of a pomegranate, given by a dodecahedron, or the planar hexagonal structure of the bee cells that has intrigued naturalists and mathematicians for centuries. Pappus of Alexandria saw in it the great wisdom of bees: in fact, with the same material used for the construction of cells, the hexagonal structure is capable of containing more honey than any others. In mathematical language, a hexagonal lattice represents the best way to divide a flat surface of equal area with the minimum total perimeter, a result rigorously proven by Toth in 1943.

The theme of the sagacity of bees has been variously taken up over the centuries. In the *Thousand and One Nights*, it is a bee that says that "Euclid could learn a lot by studying the geometry of my cells." The three-dimensional version of the problem is equivalent to the problem of the optimal packing of spheres to which the Kepler conjecture is applied. A fascinating study of the various mathematical aspects of the morphology of living forms was made by Thomas D'Arcy Thompson in the book *Growth and Form*, which has become a classic of scientific literature.

At that time, he wrote: "It is difficult to put into words the intense pleasure that I gained from this discovery. I have never regretted all the time I spent thinking about it. I was never tired of working on it, I made calculations days and nights, until I saw that my hypothesis corresponded to the orbits foreseen by Copernicus." Although suggestive, Kepler's hypothesis was clearly meaningless, if only because there are actually eight large planets and a flood of asteroids in the Solar System. However, the book brought him great fame, so much so that he was invited by Tycho Brahe to join him in Prague as a mathematical assistant to his astronomical observations. Never was an invitation more welcome, both for the notoriety of those who had made it and for the unfortunate problems that Kepler had faced in Graz recently because of his Protestant religious faith.

Magic Prague

The court of Rudolph II of Prague was a great circus. The emperor, saturnine and obsessive, was an unbridled collector and a generous patron, passionate about science and alchemy: a collection of great artists, charlatans, and half-notch adventurers all flourished under his wing, but so did great scientists, such as Brahe and Kepler. The two formed a formidable couple: if Tycho Brahe was a superb astronomer but a bad mathematician, Johannes Kepler was instead a bad observer but a superb mathematician. Furthermore, both of them were driven by a great enthusiasm and a strong passion for their work. Their actual collaboration, however, lasted very briefly, because Brahe died shortly after Kepler had arrived in Prague. Kepler inherited Brahe's monumental *Rudolfine Tables*, compiled by the Danish astronomer throughout his life, a very precise series of data on the positions of the various planets. The problem was finding a way to manage that hodgepodge of measures, i.e., to see whether or not there sense existed in that huge amount of data. Finding the end of that skein was the greatest scientific achievement of seventeenth-century science: the first two astronomical laws that bear Kepler's name were enunciated in the treatise *Astronomia nova,* published in 1609, while the third law first appeared in the *Harmonices mundi* of 1619. The three Kepler laws can be summarized as follows:

1. The orbit described by each planet in its revolutionary motion is an ellipse of which the Sun occupies one of the two foci.
2. During the movement of the planet, the ray that joins the center of the Planet to the center of the Sun (vector ray) describes equal areas in equal times.
3. The square of the revolution period of a planet is proportional to the cube of its average distance from the Sun.

A Trial for Witchcraft

This great discovery about the laws of the cosmos unfortunately did not bring any luck to Kepler. The last years of his life were mired in economic worries of all kinds and domestic miseries. In 1612, his patron, King Rudolph, died, and he thus lost his job at the court. Shortly afterwards, his wife died as well. He decided to accept the offer of a professorial position at the University of Linz and to compensate for the modest salary that was offered to him with the publication of an astrological almanac, as well as keeping up his side business of predicting the future. [AU: If I have interpreted this sentence incorrectly, and the true meaning was that the book was both an astrological almanac and a guide to predicting the future, the last clause should read '…an astrological almanac that also served as an instructional manual for predicting the future.'] He decided to look around to find a new wife and, to do this, he turned to science, specifically, to statistics! Of the eleven possible candidates, he carefully weighed the merits and demerits of each one, trying to understand what the optimal output of his research could be. He was not looking for a woman with a dowry, but rather someone who would assure him of a certain familial serenity and—why not?—also happiness. And so it was, in fact: he married the poorest of all of the "applicants," and it was a happy choice. The contentment of his new marriage was troubled, however, by the dramatic news of the arrest of his mother, on charges of being a witch. Witchcraft trials were often on the agenda at the time, and the probability that they would end badly was very high. Kepler had to rush to Würtemberg and get busy trying to exonerate his old mother from that insidious indictment. He did it by himself, transposing his propensity for logical reasoning and stringent data analysis into the legal field: he wrote a defensive memory in which he dismantled, one after the other, all of the testimonies collected by the prosecution, highlighting their contradictions and, ultimately, the fragility of the accusation. The whole file was passed to the Tübigen court, which, recognizing the poor reliability of the testimonies, exonerated the woman, proclaiming her as having been "acquitted for lack of evidence." Poor Kepler died in 1630, at the beginning of his sixties, broken by the troubles that had plagued his life and reduced to poverty as a result of having taken charge of the publication of the *Rudolfine Tables* at his own expense.

A Famous Geometric Conjecture

So far, we have talked about what is certainly Kepler's most important scientific legacy, the turning point for the birth of modern astronomy. However, his name is also associated with a famous geometric conjecture. It concerns a mathematical problem that, despite its simplicity, has required almost four centuries of effort to effect its solution, with the involvement of some of the greatest names in science, namely, Newton, Lagrange, Gauss and Hilbert. The problem originates from the question:

what is the most compact way of arranging spherical objects of equal radius in three dimensions? That is, what is the configuration of maximum density?

The problem apparently arose with Sir Walter Raleigh, the well-known English adventurer at the service of Queen Elizabeth, who needed to determine the best way to stack cannonballs in the holds of ships. Initially posed to the mathematician Thomas Harriot, who guessed the answer, the question was then turned by Harriot over to Kepler. The solution proposed by Kepler can be found in one of his books published in 1611, *Strena sue de nive sexangula*, a treatise intended to influence crystallography in the following two centuries. The book was mainly devoted to the explanation of the hexagonal symmetry presented by snowflakes and the development of a mathematical theory for the genesis of organic and inorganic forms. In the course of his arguments, Kepler posited that the optimal packing of spheres is obtained with a hexagonal stacking, that is, the one typically used by greengrocers to display oranges: to a first horizontal layer, in which each sphere is surrounded by six others, add the next layer by placing the spheres inside of the dimples of the first layer, and continue in this way for the subsequent layers. Using elementary notions of geometry, it is easy to see that, in the solution proposed by Kepler, each internal sphere is surrounded by another 12 and that the fraction of the total volume, occupied by this stacking (known in crystallography by the name of a cubic lattice with a centered face), is equal to

$$d = \pi / \sqrt{18} = 74.048.\%$$

Kepler conjectured that this was the maximum density obtainable with both regular and irregular spherical arrangements, a result known since then as Kepler's conjecture. While, on one hand, this conclusion seems intuitively obvious, its proof has been surprisingly elusive over the years, so much so that it was included on the list of the 23 most important open problems in mathematics in Hilbert's famous programme presented at the conference of the international congress of mathematicians held in Paris in 1900. Hilbert's 18th problem, in fact, states: "I would like to draw attention to this question that can be relevant for number theory and also useful in physics and chemistry: how can one arrange an infinite number of solids of the same shape as densely as possible in a space, for example, spheres of a given diameter or regular polyhedra with given sides, or in such a way that the ratio between the occupied and empty space is as high as possible?" The first significant progress was made in 1831 by Gauss, who proved that the centered face cubic lattice is the only regular arrangement that has the maximum possible density, that is, any other configuration that falsifies the Kepler conjecture must necessarily be of an irregular type. It is not difficult to find packing of spheres with a density that is locally larger than the hexagonal one, but it inevitably decreases when considering larger volumes. For example, randomly filling a box with marbles (i.e., arbitrarily shaking the container as you fill it) results in an average density equal to 64% of the total volume. The fact that it seems impossible to construct irregular configurations of spheres with densities greater than Kepler's is a clear indication of the validity of his conjecture, although it does not constitute proof of this.

Nature of Mathematical Proofs

Laszlo Fejes Toth, a Hungarian mathematician, was responsible for a further step in solving this problem: in 1953, he proved that, to determine the maximum possible density of irregular configurations, it is necessary to carry out an exhaustive analysis of a finite, although extremely large, number of cases, analysis that can be performed, for example, by means of a computer. This was, in fact, the approach subsequently followed by Thomas Hales, who, in a work started in 1992 and partially completed in 2003, applied linear programming methods to analyze the approximately 100,000 equations (in a number of unknowns ranging between 100 and 200) that describe the approximately 5000 irregular configurations identified by Toth. None of the configurations analyzed in Hales' work, published in the Annals of Mathematics in 2005, has a packing density greater than that of Kepler and, in the opinion of the college of 12 auditors appointed by the magazine, "Kepler's conjecture can be considered at the moment to be 99% valid." However, further efforts are needed for its completion.

Similarly to the Four Colour Theorem, which states that, to distinguish each nation on a geographical map, it is sufficient to use only four different colours (another famous problem solved in 1976 with the crucial aid of a computer), the proof of Kepler's conjecture pursued by Hales helped to further animate the debate on the nature of mathematical proofs. In fact, according to many mathematicians, the demonstrations obtained with the help of the computer must be considered as only preliminary stages along the way to a definitive test, which must instead be found in the logical, concise and elegant passages of the Book, the one in which, according to Erdös, God wrote all mathematical truths. It may be, however, that such a proof does not exist for some problems, in which case a new and interesting type of theorem would have appeared, a theorem that highlights the theoretical limits of mathematics. That it was the mystic Kepler who undermined this belief about the harmony of mathematics seems to be another of his myriad paradoxes.

Landau. The Ten Commandments

At the end of a conference by Albert Einstein, held in Lipsia in 1930 in front of the German Physical Society, the president, after warmly thanking the famous guest for his speech, looked around to welcome the questions of those present. From the last row of the room, a young man stood up. He appeared clumsy, with a bony face and a large rebel tuft of hair, with two very lively eyes. His German pronunciation was not the best, but what he said sounded more or less like this: "What Professor Einstein said was not stupid, however, the last equation he wrote does not derive from what he previously discussed. In fact, it requires further assumptions, but, worse, it does not satisfy an invariance criterion, as it should." Everyone in the room turned back to look at the strange goblin in disbelief, all but Einstein. Instead, he stared thoughtfully at the blackboard and, after a long and embarrassing silence, said, "The observation of that young man over there is perfectly correct. I therefore ask you to forget everything I said to you today." The young man in question was Lev Davidovicˇ Landau.

Le Rouge Et Le Noir

Lev Landau was born on January 22, 1908, in Baku, on the Casio Sea, the current capital of Azerbaijan, to a Russian family of Jewish origin. His father was an oil engineer, his mother a teacher. It quickly became clear that the boy had a certain facility for mathematics and other exact sciences. "I wasn't a prodigy," he said years later, "at school, my homework was considered 'good', but nothing more"; but he immediately added, with not without a hint of malice, "It's true though, I liked math and I don't remember when I learned to calculate a derivative or to make an integral."

The years of his youth coincided with one of the most turbulent periods in the history of Russia. In February 1917, the growing shortage of food, combined with rising inflation and military reversals on the front of the Great War, from which came news of an impressive number of deaths, had led thousands of people to go on strike in the squares of St. Petersburg, then the capital; from the Winter Palace,

© Springer Nature Switzerland AG 2020
G. Mussardo, *The ABC's of Science*,
https://doi.org/10.1007/978-3-030-55169-8_12

Tsar Nicholas II ordered the soldiers to open fire and immediately end the riots. The revolution began at the precise moment when the soldiers refused to obey and openly joined the side of the demonstrators. In the wake of the social uprising that was growing by the hour, many Russian exiles believed that it was the right time to return to the motherland, among them Vladimir Il'icˇ Ul'janov, better known as Lenin. He arrived from Zurich by train, together with about thirty other Bolsheviks, in a leaded train car: this had been the condition imposed by Germany for Lenin's passage through its territory. When the train entered the St. Petersburg station on April 3, a huge crowd awaited him. Convulsive months followed, in which, throughout the nation's cities, there was a bloody struggle between the various factions in the field—monarchists, government forces, social democrats, Bolsheviks—resulting in total chaos, with land occupations by farmers, strikes by the workers, sabotage of the railway lines, gunfights and fierce repression: in the end, the Bolsheviks prevailed and, when, on October 25, 1917, Lenin proclaimed that the revolutionaries had occupied the Winter Palace and the government was on the run, there were shouts of jubilation from the crowd of workers, soldiers and sailors of the Baltic fleet who came to the capital to help the Red Guards. The October Revolution had begun. Immediately afterwards, the Council of People's Commissars took office and issued a series of decrees to impose the dictatorship of the proletariat as soon as possible. The tsar and all of his family were killed in July 1918, but this did not prevent the outbreak of a new bloody civil war, which set the Reds, or the Bolsheviks, in opposition to the Whites, those nostalgic for the Tsarist monarchy. After countless murders and brutalities on both sides, the Reds managed to prevail, thanks to the Red Army organized by Trotsky, who liquidated the counterrevolutionary forces and established their power throughout the territory of the Soviet Union. Of the surviving whites, many preferred to repair abroad, among them, the future author of *Lolita*, Vladimir Nabokov, the brilliant chess player Alexandre Alekhine, the famous composer and conductor Igor Stravinsky, and the famous pianist Sergei Rachmaninov.

Baku had suffered several dramatic changes: first, the town had fallen under the control of the Reds, after which it had been reconquered by the Whites before definitively returning into the hands of the Reds. As a result of these tragic events, the middle school attended by Landau at the time was closed, and the little boy, who spent most of his time at home, developed an apathy and an indolence for which he received a severe reprimand from his parents. He awoke from that torpor into which he had fallen thanks to a reading of *Le Rouge et le Noir* by Stendhal and his identification with the novel's protagonist, Julien Sorel, a young man always fighting against adversity, with the dream of a heroic life.

He entered Baku University in 1922, when he was just sixteen years old. In 1924, he moved to the Faculty of Physics of St. Petersburg, which, in the meantime, had become Leningrad. There, he learned of the impetuous developments that were shaking the world of theoretical physics in those years. He plunged with all of his youthful ardor into the study of Schrödinger and Heisenberg's works on quantum mechanics; he discovered the incredible beauty of general relativity ("a feeling," he said, "that every true physicist must have"); he was intoxicated by his ability to

understand things that the human mind cannot visualize, such as the curvature of space–time or the uncertainty principle. His classmates in those years were Dmitry Ivanenko, nicknamed "Dimus," and the famous Georgij Gamow, called "Johnny"; as for him, the nickname "Dau" was chosen. Together, they formed "the three musketeers." Student life in Leningrad was quite lively, and the three musketeers often threw themselves into endless discussions on science, politics, and life itself, but they also engaged in hilarious imitations of their professors, talked about girls and gallant appointments, and poked fun at each other. In regard to everything, Landau had to propose his "theory," his own taxonomy: he had to classify which book was the most interesting and which the most boring, who was the most beautiful girl and who was the most unpleasant, who was the strongest chess player and who was the weakest, the most beautiful and the most banal, deciding whether Einstein or Heisenberg was better ... That goliardic period left an indelible mark on him, strengthening his personality to the point of making him arrogant, sarcastic, merciless, controversial and iconoclastic, or, at least, to seem to be such. His famous work on the density matrix, a key concept of quantum mechanics, destined to become the first of his "ten commandments," dates from that period.

The Mecca of Physics

1929 was the turning point: in fact, he made his first trip abroad and came into contact with the elite of European physics. In England, he visited Cambridge to see Paul Dirac, for whom he had great admiration. There is no direct evidence of their encounters, but it is not difficult to imagine the surreal scene that took place in the hushed environments of St. John's College: Landau speaking excitedly the whole time, waving his head and moving his arms continuously, Dirac looking at him without moving a muscle, occasionally answering in monosyllables - "Yes," "Yes," "Oh, no!"—to the solicitations of his guest. While in Cambridge, Landau also got to know a Russian physicist who worked at the college's Laboratory: people had told him that this man was a wizard of cold, an explorer of those very low temperatures that can be reached with liquid helium. As he discovered years later, it would turn out to have been one of the crucial meetings of his life: the man's name was Pyotr Kapitsa.

Within eighteen months, he also visited Switzerland, Germany, Belgium, and Holland, but he spent most of his time in Copenhagen—the Mecca of physics at the time—in the company of the man who he forever considered his primary teacher, Niels Bohr. In the circle inhabited by Bohr and the great physicists who gravitated around him, Pauli, Heisenberg, Ehrenfest, and many others, Landau was immediately noticed for the ability that he showed in solving any problem with which he was presented: he did it with an impressive speed, often in an original way, thanks to his total mastery of mathematics. There was no geometry or complex analysis theorem, differential equation, algebra or combinatorial theory that he did not know in detail and that he could not handle like a conjurer with his cards. Apart from the endless and lively debates with Bohr on quantum mechanics, another recurrent topic of conversation was socialism, of which he was a passionate supporter. Hendrik Casimir, years later, still remembered, in exhausted terms, the excited discussions with Dau on the Russian Revolution, and on his profound conviction that it had brought individual freedom and that socialism would only improve the human condition more and more. To emphasize his revolutionary spirit, he often went around in a bright red shirt.

Box: The ten Landau commandments

Landau's major discoveries in physics are summarized in what are known as the "ten commandments":

(1)   The density matrix in quantum mechanics and statistical physics (1927)
(2)   The quantum theory of free electron diamagnetism (1930)
(3)   The theory of phase transitions (1936–1937)
(4)   Ferromagnetic domains and the interpretation of antiferromagnetism (1935)
(5)   The theory of the intermediate state of superconductors (1943)
(6)   The statistical theory of atomic nuclei (1937)
(7)   The quantum theory of helium superfluidity (1940–1941)
(8)   The cancellation of the electric charge of the elementary particles at great distances (1954)
(9)   Fermi's quantum liquid theory (1956)
(10) Parity violation in weak interactions (1957).

In Copenhagen, Landau was in seventh heaven. He and Bohr had a great understanding of each other, even though Bohr was twice his age. Their conversations were always vibrant and lively, each one frantically trying to convince the other of their theses. It is interesting to note the description of Landau given by the wife of the great Danish physicist, Margareth: "Niels admired him and became attached to him from the first day; he understood his nature. He was - how should I say it? - unbearable; he used to interrupt Niels constantly, to laugh in the face of his older colleagues, he was just like a rascal, a brat, an enfant terrible. But how talented he was, and how honest he was!" Ehrenfest, who knew both very well, said that "a dispute between Bohr and Landau is always a show, regardless of the subject."

The only regret that Landau had was that "he had arrived too late," namely, when all of the fundamental problems of quantum mechanics had already been addressed and clarified: "All the cute girls are now married and all the interesting problems have been resolved. I'm not sure I like the ones left." In 1930, in fact, the principles of quantum mechanics had been established once and for all: the equivalence between the wave formulation of Schrödinger and that of the Heisenberg matrices, the uncertainty principle, the Pauli exclusion principle, and the probabilistic interpretation of the wave function provided by Max Born. Nonetheless, Landau found a way to occupy the rest of his life by solving many equally fundamental problems: the Landau levels, the Landau diamagnetism, the Landau spectrum, and the Ginzburg–Landau theory are just a few of a long list of theoretical successes associated indelibly with its name. On the occasion of his fiftieth birthday, his friends made him a gift of a marble plaque upon which "the ten commandments," or ten of his most important works, were carved.

Be Careful, He Bites!

Returning to Russia after that long foreign tour, he quickly made himself known for his ingenious, extravagant and hypercritical character, about everyone and everything. He was called to serve as a professor at the University of Char'kov in Ukraine. There, he met his future wife, Kora, and there he began his legend. He achieved a series of brilliant discoveries, one more fundamental than the next, concerning a broad spectrum of issues: the diffusion of the phonons in particular atomic collisions; the emission of radiation by slowed-down electrons; the theory of magnetism; the thermodynamic theory of phase transitions; nuclear reactions within a star. Needless to say, he had absolute control over all fields of physics.

He also organized conferences that attracted physicists from Moscow, Leningrad, Tbilisi, Armenia and Azerbaijan: everyone came to Char'kov just to talk to him, hear his opinion and draw inspiration from it. He established a series of fruitful collaborations between theoretical and experimental physicists. Once he began to realize the very poor scientific preparation of the students who flocked to the university classrooms, he thought of writing physics manuals at all levels: that embryonic idea, which really took shape only years later, led to the publication of the ten monumental volumes of his Course in theoretical physics. This is the text upon which generations of physicists around the world were trained, an encyclopedic work with an unmistakable style, with thousands and thousands of pages born out of the straightforward writing of a physicist who never wasted time discussing the frills, but immediately went to the core of the problem. It was futile to seek rigorous evidence of statements that were, in his opinion, "obvious"; and to those who protested that they weren't quite so obvious, he replied, in a Luciferian way, "Well, if they aren't obvious to you, then you shouldn't be a physicist."

According to the famous physicist David Mermin, "Landau's complete work arouses the same admiration as for Shakespeare's complete work or for the Köchel catalog of Mozart's work. To think that all this was done by one man is beyond imagination." His was a unique, universal knowledge, beyond any level of blind specialization. Like Enrico Fermi, for Landau, there were, in fact, no padlocks or closed doors in that vast building that is physics: he had an absolute command of all fields, and he moved with great skill among the laws of hydrodynamics, astrophysics, thermodynamics, quantum field theory, general relativity, solid state theory, low temperature physics and cosmic ray physics. Everything fascinated and intrigued him; everything seemed to exist to challenge his intelligence and his ability to penetrate the secrets of nature. And, besides being an extraordinary scientist, Landau was also a charismatic master: his lessons were exceptionally deep and brilliant, characterized by the clarity and rigor of the topics.

He had a very particular style of work: if he wasn't lying on the sofa thinking, he was immersed in heated discussions with his collaborators or students. Not an easy experience for the unfortunate chap who was ill-prepared to face him: in those moments, in fact, he could be vehement, witty and ironic, but often he was harsh, caustic and sharp as a blade. The high standard of scientific culture that he claimed for himself and required from others did not bow to anything, much less easy compromises. The word "authority" - represented by the president of the Academy of Sciences, a powerful politician or a famous scientist - had no meaning for him. The only authority that he recognized was that of logic, talent and speed of thought. He loved to talk to everyone, but only as long as his interlocutor avoided empty or specious reasoning, which he absolutely detested. And it was in these circumstances that his sharpest language appeared: with his dialectic, he mercilessly ridiculed topics that were without content or pervaded by brainy reasoning. His judgment was famous: "This work contains many new things and many interesting things. Unfortunately, the new things are not interesting and the interesting things are not new." Another of his famous quips was, "Cosmologists are often wrong, but never in doubt." Sometimes, after a conversation that ended badly, he found himself standing in the center of the room, with a doubtful air and disheveled hair, perplexedly asking himself, "But why did he get so angry? I didn't say he's a fool, I just told him that his work was complete nonsense." At the door to his office, there was a sign that warned the visitor: "Lev Landau, be careful, he bites!".

Behind all of the acerbity, however, there was great scientific impartiality, and this, combined with his unique talent, made him deserving of the position of supreme referee between colleagues and students. Not everyone in Char'kov, however, liked his hasty ways and his continual ridicule of ignorant people: the atmosphere in the Institute soon began to get heavy, and the students began to go to the party headquarters to declaim the inflexibility of that professor, especially in exam sessions. Added to this discontent were the rumors that depicted Landau as an "idealist" or, worse, an "anti-Soviet Trotskyist," fueled by the fact that he did nothing to hide his mocking comments about the various measures taken by the Soviets. One day, the rector, without provocation, threatened to fire him if he did not stop criticizing the workers' party and treating the students contemptuously: he had not even finished talking when Landau got up and left his office, slamming the door, never to come back.

Lubyanka's Guest

In early 1937, Kapitsa invited Landau to take up the position of director of the theoretical division of the Institute of Physical Problems, a newly created scientific institute in Moscow "tailored" for Kapitsa.

We have to take a step back at this point and say something about Pyotr Kapitsa: his life story had a dramatic start and a Kafkaesque flavor: in 1920, after the death of most of his family members in a scarlet fever epidemic, including his wife and two daughters, to escape the pain, Kapitsa decided to go to England, looking for a better future. He had stayed there for several years, much appreciated by Ernest Rutherford, who, in fact, strongly desired that he join the Cavendish Laboratory at Cambridge, of which Rutherford was director. In that laboratory, he had installed first-rate experimental equipment, achieving very important results in the field of electromagnetism; afterwards, he changed his research area, beginning his study on the physics of very low temperatures, immediately becoming one of the best experimental physicists in the field. His fame had grown enormously in England and abroad. By 1926, he had begun to return to Russia for short periods just to visit his relatives; during one of these stays, he met Anna Krylova, ten years younger than he, whom he married in 1927. The couple had two children and rented a larger house in the Cambridge countryside, near the cricket fields beyond the Cam River. In August 1934, Kapitsa and his wife left their children in Cambridge at a friend's house for what they thought would be a short holiday in Russia: after attending a congress in honor of Mendeleev and staying in Leningrad, they visited his wife's family, took a trip along the Volga, and enjoyed the white nights walking along the Nevsky Prospect. When it came time to leave, however, a great surprise awaited them: their expatriation permit had been revoked and they could not cross the borders of the Soviet Union! They had become prisoners in their own country... Stalin had long known of the value of Kapitsa and did not want a foreign nation like England to take advantage of that talent: he had therefore decided that the scientist should remain in his own country. A tight three-way deal began, Kapitsa on one side, the Soviet government on the other, and Rutherford offering support from England. After a grueling two-year negotiation, an agreement was reached that would allow Anna to return to England to retrieve the children and bring them to Russia, provided that Kapitsa accepted the post of director of a new Physics Institute created especially for him, with the commitment that he would never go abroad again. Rutherford, for his part, allegedly shipped the equipment left in Cambridge to Russia.

This entire latter series of events took place in 1936. It was in 1937 that Kapitsa asked Landau to join him, and the proposal was enthusiastically accepted. Even in Moscow, however, Landau was unable to keep his exuberance at bay: having a sharp tongue certainly does not help in making friends, especially if they respond to the name of Stalin, Iosif Vissarivich Dzhugashvili. On April 28, 1930, at the first light of dawn, a black van pulled up to the door of Landau's house. Four agents of the NKVD, the notorious Soviet secret police, came out and beat forcefully on the door with the butt of their rifles. When the lanky occupant opened, they took him, violently pushed him into the passenger compartment and sped off towards the Lubyanka, a gloomy yellow brick building in the center of Moscow. The sinister fame of that place was linked to the thousands and thousands of people who had been brought there in handcuffs, some of them never to be seen again. Dau was thrown into jail on charges of anti-Soviet activity and spent nearly a year in the Lubyanka's cold, inhospitable cells.

Speaking of cold, it was the "wizard of cold" who got him out of there: his old friend Kapitsa personally wrote to Berija, the fierce chief of Stalin's secret police, begging him to forgive Landau's famously bad temperament. With great courage, he added:

> Obviously, it is not for me to meddle in the affairs of the NKDV, but, even after a year in prison, no accusation has been made against Lev Landau; I, who am the director of the Institute of Physical Problems and his direct superior, do not know what he is accused of; he is in poor health and could die at any moment; his talent has been denied to the glory of the Soviet Union for more than a year. I have recently made great progress in the field of physics and Landau is the only one able to understand the fundamental experiments on superfluidity that I am carrying out in my laboratory.

Incredibly, the letter had the intended effect: Landau was released on April 29, 1939, under the personal responsibility of Kapitsa. He came out of prison as white as a corpse, thin as a skeleton, but alive. Over the next six months, Landau developed the complete theory of superfluidity that the Nobel Prize should have earned him in 1962. At very low temperatures, it is well known, the warmth of a friendship is perhaps the most beautiful thing.

The Classification of Physicists

What is the geniality of a scientist and how can his talent be measured? These are subtle questions whose answers are not at all univocal, but, for Landau, classifying or drawing up a ranking was a real art: he addressed the question by proposing two slightly different methods.

The first way to classify physicists, according to Landau, was to use five classes on a logarithmic scale. In the highest class, there was only Einstein, while the second class was led by Bohr, followed closely by Heisenberg, Dirac, Fermi, Schrödinger, etc. Landau originally placed himself halfway between the second and third classes, with a promotion many years later to the second class. The logarithmic scale of this ranking meant that a scientist of a certain class was ten times inferior to one who belonged to a higher class. As in the groups from Dante's Inferno, Landau placed in the lower class all those whom he referred to as "pathological scientists," and then gradually climbed towards the Empire, the place of the elect.

The second way, created partly for fun and partly to irritate colleagues, was a divertissement: it consisted of classifying scientists according to a series of diagrams. These diagrams were constructed from two parallel horizontal lines: the length of the lower line was a measure of perseverance, tenacity and energy, and therefore the longer the line, the greater the determination of the scientist in question; the length of the upper line, on the other hand, was a measure of originality and genius, with the assumption that the shorter the line, the more these qualities were enhanced; exceptional originality or intuition had to correspond to a point. To complete the diagram, the two lines had to be connected. In this way, according to Landau, there were four classes of diagrams for classifying scientists.

The first class included extremely original, brilliant and independent scientists, who also had a great deal of energy and care in completing their studies. This class, which obviously included the most productive and influential scientists, such as Einstein, Fermi, Dirac and Heisenberg, was represented by a triangle-shaped diagram:

The second class corresponded to the following diagram:

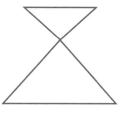

This one represented scientists who were full of energy, but not very original: scientists who disliked intellectual challenges, whose devotion to physics was somehow greater than their ability. Mischievously, Landau added that almost all Russian physicists fell into this category (no surprise that he had so many enemies at home).

The third class was, in fact, somewhat special, and consisted of all of those extremely bright and intelligent scientists who were quick to understand the essential points in any complicated situation, but who had somewhat limited energy: those who did not like to investigate all of the details and had little or no perseverance, so that many of their researches were destined to remain incomplete. The plot was therefore.

Landau said that he considered himself as belonging to this category, together with Oppenheimer, whose understanding of things was almost instantaneous, but whose originality was only marginally supported by technique and determination.

Finally, the fourth class included the lion's share of the world's physicists, with a diagram given by the following figure:

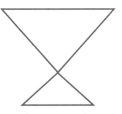

These were scientists who were neither original nor brilliant, characterized by poor determination: they had probably done something significant in the past, but they certainly would not do anything important in the future.

It worth mentioning that Landau abhorred writing, and that, for the transcription of his works or letters, he used either his faithful secretary, Nina, or his most trusted collaborators, including the workaholic Evgeny Lifšic. There was a merciless joke attached to this: it was said that, of the ten superb volumes of the theoretical physics course, of which Lifšic was among the co-authors, Landau had not written a single word and Lifšic had not contributed a single idea. Once, when Landau was asked who, in his opinion, the best physicist was, he replied, "I don't know who the first is, but I certainly know who the second is: Lifšic!".

Landau's style was based on blinding precision and impeccable logic, according to which he neglected to consider all of the trappings that were useless for the purpose of discussion. Only when all of this had been done - and only then! - did the time come for him to confront the experiments. "This saves time, and life is short," he said, a phrase that proved tragically true in his case. A road accident on the outskirts of Moscow in 1962 left him hovering between life and death for several weeks, and though he survived, the following six years were marked by slow agony. He died on April 1, 1968. Knowing his burlesque character, many people, upon hearing the news, thought it was one of his usual jokes ...

Further Reading

I.M. Khalatnikov, *Landau. The Physicist and the Man. Recollections of L. D. Landau* (Pergamon Press, Oxford, 1989)

Maxwell. Fiat Lux

With the exception of a poet, a war hero or a rock star, dying young is a terrible mistake. James Clerk Maxwell made this mistake by dying in 1879 at the tender age of 48, and although, for physicists, he still remains one of the most significant figures in his field, his name means little or nothing to most people. It is unlikely, in fact, that anyone knows that our modern technology is due to him: colour televisions, cell phones, digital photography and computers are all devices based on the principles of electromagnetism.

In reality, this idea points to a question that is much more general and covers the whole sphere of knowledge, that is, what do we know about the world surrounding us and the laws that regulate it? On the basis of many experiences, today, we know that electrical forces largely influence the physical and chemical properties of matter, from the atom to living cells: the water that freezes, the plant that grows, the northern lights and the grandiose spectacle of lightning are all phenomena of an exclusively electric or magnetic nature. This is encoded, as such, in just four equations: Maxwell's wonderful equations. These not only encompass the whole theory of electrical and magnetic phenomena, but also explain the laws of optics and the nature of light and all other forms of radiation, such as radio waves and X-rays. Maxwell elaborated these equations in 1861, when, after long years of study, he came up with an idea that turned out to be deeper than any book of philosophy, more beautiful than any painting, more powerful than any political act. Nothing would be as it had been before.

The Mysteries of Electricity and Magnetism

Throughout the mid-nineteenth century, the best physicists of the time had searched for the key to explaining the mysteries of electricity and magnetism. The two phenomena seemed to be inextricably linked, but the final nature of this link had

© Springer Nature Switzerland AG 2020
G. Mussardo, *The ABC's of Science*,
https://doi.org/10.1007/978-3-030-55169-8_13

proved somewhat elusive, despite all of the attempts made to decipher it. The solution to the puzzle is due to Maxwell. His theory of electromagnetism, one of nature's fundamental interactions, is a central pillars for understanding our world: it paved the way for two true triumphs of twentieth-century physics, the theory of relativity and quantum mechanics, and it survived, unscathed, the conceptual jolts produced by these two great revolutions. As another famous physicist, Max Planck, wrote, Maxwell's theory must be remembered as one of the most important intellectual achievements of all time.

If its author had lived only a little longer, he could have, on December 12, 1901, rejoiced with Guglielmo Marconi when the latter received the first radio signal, one of the most surprising predictions of Maxwell's equations. Or perhaps he could have developed the theory of relativity first: after all, it was precisely Maxwell's equations that inspired Einstein to demolish the concepts of absolute space and time. Furthermore, it was upon reading Maxwell's article on the ether, in the ninth edition of the Encyclopaedia Britannica, that Albert Michelson devised the optical interferometer by which, together with Edward Morley, he discovered that the speed of light is the same for all observers, regardless of their state of motion.

A Cultivated Country Gentleman

Everyone knows Isaac Newton and Albert Einstein, but, as we mentioned above, besides within a close circle of specialists, James Clerk Maxwell is practically unknown. Why this is so is a mystery, although the reason is perhaps due to his disarming modesty: he was not a scientist who elbowed his way into promoting his work and there was not even anyone who could do that for him, as in the case of his contemporary Darwin, who was also reluctant to receive the honors of the general public, but who was lucky enough to have an exceptional sponsor in Thomas Henry Huxley. Instead of wasting his time chasing after honors and prizes, Maxwell loved to sit quietly in his Scottish estate, where he used to stroll through the lakes and meadows of Glenlair, to experiment (sometimes in the company of his wife) and meditate on the results obtained or on the mathematical theories that governed them.

Anyone who met him in his years of maturity would have been fascinated by his slender figure, the sweetness of his piercing eyes, the large beard that framed his face, the timbre of his voice in which a strong and unmistakable Scottish accent stood out; they would have found themselves in front of a generous and jovial man of great charm, one of those types who likes to be in the company of others, perhaps to entertain friends with jokes or witty linguistic games, made without any base vanity. For the sober elegance of his Shetland tweeds, paired with sporty waistcoats and soft flannel shirts, he would have been mistaken for a cultured country gentleman, as existed in great numbers in the Victorian era: a gentleman certainly weaned in the college at Oxford or Cambridge, but more at ease among the green expanses of the English countryside than in the muffled atmosphere of the clubs, happy to be able

to be outdoors, go horseback riding, bathe in the lakes or stop in the pubs along the rivers, drinking a double malt whiskey. On the other hand, one would never have suspected oneself of being before one of the greatest scientists of all time, able to compete with Newton or Einstein in terms of originality and depth. A scientist of prodigious skill, capable of making everything seem so simple as to appear almost natural: in his writings, everything proceeded like the notes of a symphony, one equation after another, interspersed with extremely elegant logical arguments. However, it was not difficult to guess that, behind that extraordinary lightness, there was actually hidden an exceptional skill, that rare skill that comes from the absolute mastery of all of the grammar of the great book of Nature; a bit like Mozart in music, who knew how to transform the pages of the pentagram into a river of sounds and make the sublime quality of his works out of the gracefulness of his notes.

A Place Called Glenlair

James Clerk Maxwell was born on June 18, 1831, in Edinburgh, into a family that could boast illustrious personalities in the fields of politics, geology, music and poetry in previous generations. He spent his childhood in Galloway, in the southwest of Scotland, where his parents, John Clerk Maxwell and Frances Cay, had moved after inheriting a large manor. The place was so isolated that there were no houses or schools nearby. It was the father who personally carried out the work on the new house, which was given the name of Glenlair: James, infected by his paternal enthusiasm, became so fond of that place that he chose it as his ideal eternal refuge, both in the years of youth and for his maturity.

John Maxwell was a very sociable person, but also a somewhat eccentric type, who spent his afternoons inventing little kits and building contraptions and other devilries that young James liked very much. As a result of these, from an early age, James was fascinated by the world around him and joyfully committed to understanding how things worked. Every object that moved, sparkled or made sounds immediately attracted his curiosity: he would then ask, with his very strong Scottish accent, "What is that?", and if he was answered fleetingly that it was a bird with blue feathers, he would insist, "How d'ye know it's blue?"

Until the age of eight, Maxwell's life was a very happy and fulfilling one, marked by the nursery rhymes read by his mother in front of the fireplace and by the games played with his father with tubs in the pond, in the company of their dog Toby. However, that idyllic climate ended with Frances's premature, struck down by stomach cancer at the age of 48, the same illness that would later hit him at the exact same age. His reaction to such a tragic event was of a disarming innocence: "I am so happy that she died; at least now, my mother will no longer suffer." Later, the bond between father and son became even closer, and the boy was very happy to accompany the father when he made the rounds to inspect his property[?]. To give him an education, John Maxwell entrusted the boy to a tutor from the neighboring village, a rude man with somewhat questionable ways: he did not hesitate to beat the boy until he bled if he did not complete his Latin exercises; James' stoic silences would infuriate the man even more, leading to outbursts of rage.

That cruel treatment left lasting traces on James' character: according to his friend and first biographer, Lewis Campbell, the violence suffered as a child was the cause of hesitations when he spoke and his sometimes oblique responses. It was only the intervention of his maternal aunt that saved him from that hell: she demanded that the lessons be ceased and that the boy come to stay with her in Edinburgh.

There, his debut in society was decidedly tragicomic: on the first day of school, he presented himself wearing a strange dress sewed by his father, a mishmash of tartan fabrics of various colors, put on as they were. The very strong Scottish accent, combined with that ridiculous country dress, did the rest: he was immediately baptized by his companions "Dafty," a name the meaning of which will be immediately evident to speakers of UK English, but that, for Americans, might be translated as "Freak or "Weirdo". Not surprisingly, they not only enjoyed mocking him, they also tore his bizarre Harlequin costume apart. But, rather than act scared, young James's reaction was instead one of wonder, as if it were a new game that he wasn't used to. Because he had inadvertently robbed the bullies of their desired irritation and upset, the teasing quickly diminished: after a few days, his new comrades, knowing him better, started to love him very much. Among them were Lewis Campbell and Peter Guthrie Tait, with whom he remained friends throughout his life.

He often wrote curious letters to his father, stuffed with jokes, word games and comic absurdities:

> Dear Mr. Maxwell, Today, I saw your son, who told me that he has not been able to solve his riddle. However, he spoke with a simpleton who was about to drown in a pool of water and who, as soon as he came out, swore not to touch the water anymore until he learned to swim, dear Jas. Alex. M'Merkwell, letter mailed who knows where.

Subsequent university studies certainly did not discourage his unconventional behavior: at the age of twenty, when he was at Trinity College in Cambridge, he used to wake up at two in the morning and spend half an hour running through all the corridors of the building, until his colleagues, awakened by the hustle and bustle, began to target him, throwing shoes at him from their rooms. He had also preserved some ancient Scottish customs, such as diving into a pond, first face down and then back swimming vigorously for about ten minutes: he said that all this stimulated circulation!

The Fingers, The Head and The Heart

As a young man, in a letter to a cousin, he wrote that, in order to learn, three qualities are needed: fingers, head and heart, and, as a matter of fact, Maxwell was a champion in all three. As a student, he demonstrated the skill of his fingers (metaphorically speaking, that of his memory and his technique), brilliantly passing the Tripos, the fearsome exams of Trinity College; as an adult, he instead used his head to set up his theories, combining sophisticated physics reasoning with elegant mathematical models. The main source of his genius, however, was his heart, namely, that profound intuition of the laws of nature that led him to discover its deepest secrets. The imprint left by Maxwell on physics is indelible: his place in the history of physics is ensured by his revolutionary research on Saturn's rings, the viscosity of gases, dimensional analysis, thermodynamics, optics, statistical statistics and, of course, electromagnetism. He was also responsible for the formulation of the principles underlying modern color theory, to which he successfully dedicated many years of study: Maxwell actually created the science of quantitative colorimetry, devising the "color triangle" (a diagram still used today in photography) and the "color box" (an extremely useful tool for studying composition). By means of a spinning top that, rotating, combined red, yellow and blue, he came to understand how the human eye recognizes the various colors and identified the receptors responsible for partial or total blindness. He successfully applied the principles of colorometry that he discovered to create, among other things, the first color photograph, the subject of which was a flamboyant Scottish tartan strap.

The Calculation of Probabilities

To explain the laws of thermodynamics and those of gases, he had the brilliant idea of introducing the probabilistic methods of statistical mechanics. The great problem of the physicists of the nineteenth century consisted in having to explain not only the laws—long known—that relate the volume, pressure and temperature of a gas, but also the general principles of thermodynamics, discovered in the first half of the century and proven crucial for both pure and applied science.

Maxwell's contribution to the solution of this problem was of extraordinary importance: starting from the idea that it was implausible that all of the molecules of a gas had the same velocity, he introduced the hypothesis that the velocities of the molecules differed, both in direction and in size, and that the whole problem was therefore purely probabilistic in nature. With these premises, Maxwell quickly came to a very elegant demonstration of the probabilistic law that regulates the distribution of the velocities of the molecules in a gas, a result later taken up and further refined by Boltzmann (see the chapter dedicated to him). Thus, he came to explain the nature of a gas's temperature and its link with the average speed of the molecules: the higher the latter, the higher the temperature. He also predicted that the viscosity of a gas does not depend on the pressure to which it is subjected, but only on the temperature, a decidedly counterintuitive hypothesis, but one that he took care to verify directly by conducting some experiments with the help of his wife.

Maxwell's Demon

Thanks to Maxwell, the calculation of probabilities was therefore fully included among the essential components of the mathematical apparatus used in physics.

However, the Scottish scientist was well aware that, despite the important successes achieved, there was still much to understand, especially in the case of the second principle of thermodynamics that regulates the irreversibility of natural processes. He therefore conceived the famous "Maxwell's demon," a hypothetical microscopic creature that he invented precisely to highlight the problems that the statistical approach posed to thermodynamics.

If you accept the idea that a gas can be described as a set of particles and that their equilibrium state is actually only the most probable state of the system, nothing prevents you from considering the existence of thermodynamic fluctuations, usually only excluded for their high improbability and not for physical reasons: this was Maxwell's reasoning. He therefore imagined a fantastic being, a little devil, who, operating on the basis of well-established physical laws, was able to violate the equilibrium conditions of a gas, for example, by opening and closing valves without any friction. Apparently a perfect bureaucrat, he was capable of accomplishing what has never been seen in nature: producing a flow of heat from a cold substance to a hot one. Here is what Maxwell wrote:

> Suppose we have a container full of gas, divided into two parts, A and B, by a septum in which there is a small hole, and that a being who can see the individual molecules opens and closes this hole so as to allow only the faster molecules to pass from A to B and only the slower to pass from B to A. In this way, without doing work, he will raise the temperature in B and lower that of A, contradicting the second law of thermodynamics.

Maxwell's demon remained a logical puzzle of thermodynamics for decades, but the attempts made to overcome it were not in vain; rather, they stimulated the development of information theory, a science that is at the basis of our modern computerized civilization. The importance of this imaginative creature has also crossed the boundaries of the area in which it was originally invented, ending up becoming an emblematic example in physics of the Gedankenexperiment: a mental experiment, a particularly effective way of carrying out scientific arguments without the use of real experiments, a technique later widely used by Einstein to explain the principles of general relativity.

A Field of Forces

It should also be mentioned that Maxwell's equations not only systematized all electromagnetic phenomena, but also gave the coup de grace to a particularly strange idea of the installation of Newtonian physics, that of instantaneous action at a distance. As Einstein said, "with James Clerk Maxwell ended one scientific season and began another." The fact is that, in the new scheme of Maxwellian physics, the interaction between two bodies no longer occurs immediately, but is transmitted by a field of forces. This force field, which acts as a perturbation of the space between the two bodies, travels at a finite speed, that of light. It is therefore only for the very high value of this speed that the interaction seems instantaneous! This is such an important observation, as we will see, that it has radically changed our vision of the physical world.

Faraday (see the chapter dedicated to him) has the merit of having introduced the concept of the "field": a continuum of forces that fill a space, such as to unequivocally determine the electrical and magnetic phenomena that take place there. However, the great experimental physicist was not adequately prepared to translate what he observed in the laboratory into a rigorous mathematical treatment, so that his contemporaries made an easy game of saying that the lines of force, highlighted by him with iron filings, were only a useful device, a way like any other to describe phenomena whose real origin was rather to be sought both in magnets and in electric charges. Maxwell, the Mozart of physics, was instead responsible for the construction of the new theoretical building, in all of its imposing architecture, the complete elaboration of field theory in the modern sense of the term.

Viewed in retrospect, the course of physics until 1820 was the triumph of Newton's scientific program. The various forces of nature—heat, light, electricity, magnetism, chemical reactions—had all been progressively reduced to the attraction or instantaneous repulsion of particles. The scenario of the Newtonian universe in which all physical phenomena took place was the three-dimensional space of classical Euclidean geometry, an absolute space, always immobile and immutable, immersed in an equally absolute time. The elements of the Newtonian world that moved in absolute space and time were the material particles that, in mathematical equations, were treated as "material points": Newton considered them small, solid and indestructible objects, of which all matter was made. The model recalled the atomism of the ancient Greeks, with the important difference that, in the Newtonian model, there was a precise description of the force acting between the material particles: in most cases, it was a very simple one, dependent on the reciprocal distance of the particles, and whose action was manifested instantly at any distance. The extraordinary success of the mechanistic conception of nature gave rise to the belief in the physicists of the early nineteenth century that the Universe was actually a huge Clock, which functioned according to Newton's laws of motion, seen as the fundamental ones of nature. Newton's mechanics was conceived as the definitive theory of natural phenomena.

However, it was not long before new discoveries made clear the limits of the Newtonian model and showed that, in reality, none of its aspects had absolute validity. The first cracks occurred with the discovery made by Hans Christian Ørsted in Copenhagen in 1820: in a famous experiment, the Danish physicist highlighted not only that an electric current is able to influence a magnetic needle, but that the force is directed orthogonally to their distance. This aroused considerable surprise among physicists: Faraday took note that this new phenomenon was not attributable to the old patterns and that new ideas were necessary for its understanding.

A Scientific Masterpiece

Maxwell's first paper on electromagnetism, *On Faraday's Lines of Forces,* saw the light in 1855, and it shaped the mathematical treatment of the electromagnetic field, which would then assume its definitive structure in the scientific masterpiece *A Treatise on Electricity and Magnetism,* finished in 1873. According to Maxwell, the notion of a line of force allowed for the elaboration of a geometric model of physical phenomena, in which the whole space was an interweaving of curves that provided information on the direction of the forces acting point by point. This geometric approach had the function of avoiding conjectures on the physical nature of electricity and magnetism, but gave importance instead to mathematical reasonings: this led to a level of precision and generality that allowed him to associate a real physical reality with the electromagnetic field.

In this way, the theory, initially developed on the basis of pure mechanical analogies, ended up crystallizing in a set of differential equations that did not need any support other than their elegance and logical coherence: it was like removing the scaffolding from the dome by Brunelleschi and seeing, with wonder, that not only did it remain standing, it rushed vertiginously towards the sky thanks solely to the harmony of its arches. Many phenomena became immediately intelligible, while new and extraordinary effects appeared on the horizon: it was a risky and, at the same time, sublime theory. In particular, according to Maxwell, there were reasons to believe that space was traversed by a disturbance that allowed for the transmission of forces from point to point. His equations made it possible to express the energy stored by the field at every point in space and to deduce the action that it exerted on the conductors, the magnets and the electrified bodies. The propagation speed of the electromagnetic field, which he calculated, was very close to that of light, and there were therefore strong reasons to say that light was nothing more than a form of electromagnetic radiation! Consistent with Maxwell's style, it was a simple and penetrating idea at the same time, with profound implications for science and technology, even if several decades had to pass before its implementation.

In his theory, there are no material actors. The equations express only the laws that govern the electromagnetic field: they do not, as in Newton's laws, instantly connect two events separated by a great distance; on the contrary, they tell us how the electromagnetic field, here and now, depends only on its value in the immediate vicinity and on the instant just passed. It is as if the electromagnetic field lived a life of its own, regulated only by Maxwell's equations. They show us the existence of a substantial symmetry between electricity and magnetism, and summarize in one fell swoop all of the laws valid in both areas.

In the words of Einstein and Infeld, "Maxwell's equations represent the most important event that occurred in physics from Newton's time onwards, and this not only because of the abundance of contents, but also because they provided the model for a new kind of law." In fact, Maxwell's theory presents itself as a theory of continuum, as opposed to the discontinuous one that had dominated the mind of researchers for about two centuries, and that had been considered the most qualified candidate for the role of universal science. Only in recent decades have the radical developments that occurred in physics, thanks to the quantum field theory originated directly from Maxwell's theory, opened the way to unification of the fundamental forces operating in nature. Maxwell's superb conceptual architecture thus ended up being one of the largest intellectual buildings of all time, erected thanks to something that he would not have hesitated to call magical.

Further Readings

B. Clegg, *Professor Maxwell's Duplicitous Demon* (Faber & Faber, London, 2019)
A. Einstein, L. Infeld, *The Evolution of Physics* (Cambridge University Press, 1878)

Nucleus. My Name Is Lise Meitner

The Diamond Ring

Everything had happened so fast—in the last few days, in the last few hours—that she hadn't even had time to catch her breath. Of course, things had started to degenerate months earlier, when Hitler had ordered the German army to invade Austria and, immediately afterwards, proclaimed its annexation by Germany: in fact, from March 12, 1938, the Austrians present in the German territory had suddenly lost their guarantee of foreign citizenship and had to submit to the same legal provisions that applied throughout the Reich, especially if they were Jews. It was for this reason that she had long since abandoned her modest apartment in central Berlin and moved to a room in the Adlon Hotel, "for greater security," as had been suggested by various friends. Together with many other senseless things, she had accepted that too: after all—she said to convince herself—the hotel was right in front of the Kaiser Wilhelm Institute of Chemistry, and therefore was a much easier place from which to reach her beloved laboratory, where she had worked tirelessly, even at night, for the past twenty years.

That day, June 13, 1938, the sunrise over Berlin was as sweet as ever: shades of pink could still be seen in the distance, the air was warm, perhaps a little sticky due to the humidity of the summer that was now knocking at the door. The night before, after spending the day at work in much the same way as she had all of the others, so as to avoid arousing suspicion, once back in the hotel, she had packed all that she had into two small suitcases. At 10:30 pm, Paul Rosbaud—an old and dear friend, who was also a chemist—picked her up and drove her to Otto Hahn's house. Once there, all three dined quickly in that large dining room that had seen much happier moments, for example, the many evenings suffused the typical Berlin atmosphere, when Otto would sing arias taken from the symphonies of Brahms or Beethoven—as he also often did in the laboratory when he was beaming over some new discovery—or when she would chat with Edith, Otto's wife, about the most recent theater show or their last trip to Potsdam, while the Hahns' little Hanno scampered about under the piano near the window. The night before, however, there had been only tense

© Springer Nature Switzerland AG 2020
G. Mussardo, *The ABC's of Science*,
https://doi.org/10.1007/978-3-030-55169-8_14

faces and sad eyes full of fear. It was a dark time for everyone: the new director of the Institute of Chemistry was a member of the Nazi party and the black uniforms of the notorious Schutzstaffeln were now everywhere. With increasing frequency, they would pull up to households in their trucks, break through the doors and carry away women, men and children to destinations that nobody knew or dared imagine. The huge glass chandelier in the dining room that evening emitted a warm light, and Lise, looking up from the table, saw Otto Hahn's beautiful face furrowed by ever deeper wrinkles: when she asked him about his wife—she was in the psychiatric clinic after another nervous crisis due to the tension of the last few weeks—he had rolled his eyes without adding further comment. Then, Otto had instructed the maid to accompany her to the guest room on the first floor: once in bed, Lise had spent a sleepless night, haunted by a thousand fears, with a dry mouth and her heart in her throat, torn by a nervous tension that the darkness of the night had done nothing but enhance beyond all limits.

In the morning, a strange thing had happened, a decidedly singular episode, which, however, she was unable to fully grasp; a part of her was uncertain as to whether it had really happened or if she had simply imagined it. While saying goodbye to Hahn, he, with eyes swollen and red with emotion, had placed the diamond ring that had belonged to his mother on her finger, with the insistence that she use it if she needed to. The ring was there, right on her finger, so there was no doubt, she could not have dreamed it. But she also seemed to remember that, in embracing her, Otto, for the first time, had not called her "Fräulein Meitner," as he had inevitably done for the thirty years since they had first met in the corridors of the University of Berlin, but instead whispered her name, Lise. It had all been very fast, strange and moving: it was just a moment and, a second later, she was sitting in Paul Rosbaud's black car, headed for Berlin Station.

The Wonder of the Old Professor

During the journey, from the window, she observed the flow of the city before her eyes with some bewilderment: the greenery and imposing trees of the large Tiergarten park, the bridges over the Spree River, the Bauhaus buildings along Nürnberger Straße, which had aroused so much curiosity when they were built in the 1920s … She followed the long avenues around the Brandenburg Gate, seeming to glimpse the golden statue of the Victory column among the thick branches of the trees. With great dismay, she noticed that long drapes displaying a hooked cross were hung everywhere, the streets were full of soldiers (on the corner of Friedrichstraße and Jägerstraße, she had counted ten or fifteen wearing the typical brown shirt of the Sturmabteilung), and in the distance, she could hear horrible choruses, the shameless screams of some squadron. Along the way, she had also noticed that several shops were closed, large yellow stars with the word 'Jude' placed in their windows. The sight of every patrol of the Kriminalpolizei encountered along the road gave her a jolt and her heart seemed to go crazy; she had been forced several times to hold her

breath, to turn her gaze forward, trying to appear as nothing more than a distinguished and harmless sixty-year-old lady in a car with her nephew, beginning a trip out of town or going for a visit to a relative in another part of Berlin. At a certain point, however, she had become panic-stricken, seized by a fear that had almost paralyzed her: at that point, she had begged the good Rosbaud to bring her back, to return to the hotel with her. She felt that she no longer had the courage to leave, she just wanted to go back to her laboratory, to her workbench, but Rosbaud, with his calm and reassuring voice, had told her that this was not possible, that it was necessary for her to leave Berlin and Germany as soon as possible, because there were chilling stories about Jews who had disappeared from their homes, their shops, their places of work.

During that long drive through the city that she had loved so much—for the music, the theaters, the people—and that now only aroused terror in her, they had also passed in front of Max Planck's house, at No. 21 Wangenheimstraße. In a flash, she remembered their first meeting, when she had just arrived in Berlin in 1907, and, in particular, the wonder of that old professor when she had expressed her intention to follow his lessons: "But you have already graduated, what do you want to learn more for?" She had tried to explain to him, timidly, that she wanted to achieve a true understanding of physics, and remembered that Planck had spoken to her in a kind, but very conventional, manner, without elaborating on the matter. It was clear that he didn't have a great opinion of women, especially if they were scientists or students …

Lise, however, was used to it: in Vienna, where she was born and where her beloved sisters still lived, she first had to overcome her father's resistance, and then that of the university authorities, to be allowed to attend the Faculty of Physics. Hers was a typical Jewish family of the upper Viennese bourgeoisie: a thousand interests were cultivated at home—music, the emergent art of photography, literature—but none of this counted: according to the laws of the Kingdom of Austria, women were prohibited access to high school studies. She therefore had to study privately and pass a very severe exam in order ultimately to enroll in the Physics Department of the University. There, she had the great opportunity to follow the exciting lessons of Ludwig Boltzmann: she always remembered, with tremendous emotion, that great man, animated by a huge passion for the truth of science, his lively discussions on atomism, thermodynamics, the irreversibility of physical processes; she found herself reflecting on how much his charisma had ended up influencing her whole life, the choices that she had made after graduation, her desire to become a woman of science, to understand the laws of nature, and to grasp the mysteries of the atomic nucleus, radioactivity and the perpetual breaking of matter. All questions born, after all, from that same great curiosity that she had felt as a child when she saw, for the first time, the iridescence of the thousand colours reflected in a simple droplet of oil. She returned in her thoughts to the great Max Planck, the father of quantum mechanics, the one who first opened the doors of that wonderful atomic world of the infinitely small: she was really moved in remembering that he had always treated her like a daughter, inviting her to his home for a dinner, a coffee or a reception with other scientists, to the point that Lise had immediately made friends with his two twin daughters, joining them on field trips and boat rides on the lakes of Potsdam. Perhaps that great man had changed his mind about women in science, "given all the support he has given me over the years, having proposed me several times to the Nobel Prize committee, having taken me on as his course assistant."

Box: Nuclear Fission

After her daring escape from Nazi Germany, Lise Meitner remained in the Netherlands for all of June and July, awaiting some future prospect. Her escape had been a coordinated effort involving many European physicists, all of whom had understood in time that the life of the great scientist was hanging by a thread. Among them were Niels Bohr, Peter Debye, Adriaan Fokker, Wander Johannes de Haas, Carl Bosch, Otto Hahn, Paul Rosbaud, and Dirk Coster.

Sometime later, she was able to see Otto Hahn again in Copenhagen, at Niels Bohr's home, and she remained in epistolary contact with him. In particular, during the Christmas holidays of 1938, she received a letter from Hahn regarding the very strange results that he and Fritz Strassmann had found in experimenting with radium, thorium, actinium and other transuranic elements. Such results could invalidate years of research on radioactivity, and the tone of the letter expressed frustration and bewilderment. Hahn hoped that Meitner could find the solution to their puzzle: "There is something VERY STRANGE about the isotopes of radium, something we don't want to mention to anyone but you. By irradiating them with neutrons, barium is among the products of nuclear reactions! Maybe you can suggest to us some fantastic explanation." To understand Hahn's bewilderment, it is necessary to know that the number of nucleons (protons and neutrons) of barium is about half that of radium and other transuranic elements: how could barium be produced by irradiating the radium with a neutron?! Understanding this oddity became a great challenge for Lise Meitner.

Christmas was approaching. On December 21, 1938, Hahn and Strassmann had sent their article on that strange discovery to the newspaper "Die Naturwissenschaften," and also sent a copy to Lise Meitner, accompanied by a long letter: "How wonderful it would have been if you had been here to work together with us, you would have been very surprised by the huge number of experiments we have carried out on this abnormal presence of barium among the decays of radium." That same day, Lise Meitner received a visit from her nephew Otto Frisch, also a nuclear physicist. He wanted to talk to his aunt about some of his experiments, but he saw her completely immersed in her thoughts. Lise proposed to her nephew that they go for a walk in the snowy woods near the house. The possibility that Hahn had interpreted the experiment incorrectly was excluded; Hahn was too good of a chemist … It occurred to Lise that Bohr had repeatedly underlined that the atomic nucleus was like a drop of water: thus, could it not have been that the nucleus had split under the external stress of a neutron? The two stopped in the snow, excited. Lise pulled a notebook out of her pocket, she and her nephew sat down under a tree, and they began to calculate the energies at stake, the initial and final masses of the nuclear reaction. It took little for them to be convinced that Hahn and Strassmann had actually observed the fission of the nucleus in their experiment, with an enormous production of energy and other neutrons! In the following days, Lise Meitner and Otto Frisch wrote an article that was published in "Nature" at the beginning of the next year, and Lise informed both Hahn and Bohr of this great intuition. Bohr was leaving for the United States and, once he arrived in New York, he spoke about it at a conference; in the blink of an eye, the news was on the lips of all American physicists. This, in fact, marked the beginning of the Manhattan Project, leading to the creation of the atomic bomb.

In 1944, the Nobel Prize in Chemistry was awarded to Otto Hahn "for his discovery of nuclear fission." Lise Meitner was not considered at all and Hahn didn't even mention her name during the awards speech. Invited to participate in the Manhattan Project during the war years, Lise always refused to be part of it, believing that science should never bow to war purposes or projects that went against humanity. Until her death, which occurred at the age of eighty-nine, she committed herself to the peaceful use of nuclear fission. She died in Cambridge on October 27, 1968, in her nephew's house, where she had been living since 1960. On her grave, the inscription reads: "Lise Meitner, a physicist who never lost her humanity."

The road ran by fast; they had crossed the whole city by now. Rosbaud drove carefully and with his usual calm, inspired confidence. Just before, he had said, "Hopefully, Coster is already at the station and there will be no problems." At the sight of the neo-Renaissance facade of the Lehrter Bahnhof railway station, he finally exclaimed, "Here we are!" Parking the car, he helped her out, grabbed her suitcases and accompanied her to the entrance. As she crossed the sidewalk, Lise felt like her legs were made of jelly; she tried to appear composed, but, in actual fact, she was extremely agitated. With one hand, she adjusted her hat.

Track 12

The station's main hall was teeming with people, the loudspeaker announcing the arrival and departure of the trains for Munich, Stuttgart, Göttingen … The large glass canopy embraced twenty tracks, each with a small gate at the entrance where the railway control staff were stationed. Lise tried to locate the train for Holland, via Bremen: she saw that it was at platform 12, so she walked towards it, accompanied by Rosbaud, who remained diligently by her side carrying her suitcases, just like a nephew with a dear old aunt. The station was full of SA agents, their dogs barking, their footsteps rumbling: they stopped and inspected anyone who seemed suspicious.

She heard the excruciating sound of their whistles, then the calamitous din of an arrest. She pretended not to have seen anything, tried to appear "normal," when, in fact, nothing was normal any more, everything was in the hands of chance: in fact, she didn't even have a passport—her Austrian passport was no longer valid and the request for another one made on her behalf by the president of the Kaiser Wilhelm Society—addressed directly to the Minister of Education—had been rejected, thanks to the negative influence of the Minister of Interior of the Reich, Heinrich Himmler!

Finally, she arrived at platform 12. Rosbaud put her suitcases on the ground, hugged her tightly, wishing her a good trip, and then left, looking back from time to time to see what was happening. Lise opened her purse, took out her ticket, and passed it to the controller: the latter looked it over, his eyes darting back and forth between the ticket and her face, her face and the ticket, and for a moment, she thought that she had been found out. She was terrified at the thought that he too would now blow his whistle, the sound would fill the station, and the SA would arrive, arrest her and take her away. The seconds ticked by endlessly. The man looked up again, stared at her one last time, and finally handed the ticket back to her slowly, inviting her to pass. Down the platform a ways, behind the column of a shelter, she met Dirk Coster's gaze. Fortunately, he was there! They exchanged a nod of understanding, then each went their own way, she towards carriage 6, he towards carriage 9. The train pulled out of the station ten minutes later.

A Train Headed to Holland

The convoy, fortunately, was not very full: in the compartment before hers, she had seen another older lady. When she had passed by, the lady had given her a smile and a nod. While watching the countryside through the train window, she began to think of all the years spent in Berlin, the initial meeting with Otto Hahn, their unceasing laboratory work carried out for such a long time. When Hahn offered her the chance to join his research team at the end of 1907, the director of the Institute of Chemistry at the time had made it clear to her that women were not tolerated in the building, so he could take her on only if she set up her own laboratory in the basement, where the carpentry shop used to be, and if she needed the bathroom, she would have to use the one in a nearby hotel. She accepted.

Every now and then, the roll of the train would toss her left and right, but she continued to fall back into her thoughts, reminiscing about how she had equipped her laboratory with three electroscopes and a chemical bench, the radioactive substances that she needed displayed on shelves, in small lead cylinders. Despite all of the bitter morsels that she had had to swallow, it had been a formidable period: in the first two years of working with Otto, she had published ten articles, and countless others had followed. Their reputation as nuclear physicists had grown rapidly throughout all European scientific circles, and even overseas. Lise Meitner smiled, remembering that, when she had met Ernest Rutherford in Stockholm from in 1908, when he had come there from Berlin to collect the Nobel Prize for Chemistry, he said in amazement, "And I believed you were a man!".

Hers and Otto's reputations had further consolidated in 1918, when they discovered protactinium, a new, highly radioactive chemical element. She had done most of the work, but was happy to share the credit: "Otto had contributed only at the end, but that's okay ..." That same year, she had also finally received her first salary: Max Planck had to remind those in charge that "Lise Meitner had received a tempting offer from Prague," a statement that finally cleared the issue. She was also reminded of the first time she met Niels Bohr in Berlin in 1920, and the pleasant summer spent the following year in Copenhagen with Bohr and his family. She recalled her conversations with Einstein, who had once made her very happy with his flattering comment: "Lise Meitner is our Marie Curie!".

The train rattled on through the countryside, the German woods giving way to tulip fields. In the distance, one could see large boats going up the canals, church bell towers and houses in the Flemish style. Lise had dozed off slightly in her seat as her memories continued to surface, one after the other: the very pleasant time spent with the Planck family, or those with Gustav and Ellen Hertz, or with the poetess Hedi Born, wife of the physicist Max Born, or all of the discussions on ancient Greece that she had had with Max von Laue, when he was not busy irradiating his beloved crystals with X-rays ... She remembered the competition with the Enrico Fermi group in Rome and the equally cohesive one with the Joliot-Curies in Paris, the race to be the first to penetrate the secrets of the nucleus and its transmutations, almost following in the footsteps and dreams of the ancient alchemists in a strange way.

Abruptly, the train slowed down, the metallic screeching of the brakes accompanied by a cloud of steam that entered and overwhelmed the carriage. She awoke with a start; outside it was almost dark. Seven hours had passed since her departure from Berlin, and they were now close to Nieuweschans Station, on the border with Holland. Outside, she heard the agitated voices of the soldiers and customs officers, the officers constantly shouting their orders, the sound of the locomotive mixing with the hustle and bustle along the quay. Once the train had come to a complete stop, all of the doors opened and German soldiers and Dutch border guards came on board, yelling. Lise saw that Dirk Coster had entered her carriage, and he immediately rushed over to converse with the Dutch gendarmes, to whom he exhibited a letter, then another document, while pointing towards her with his finger. Finally, she saw that Coster was waving his hand to the Dutch officer, inviting him to speak with the German guards. Lise sat petrified as she witnessed this scene. Before leaving, Rosbaud had told her that Coster knew how to get her across the border: "You will see, the Dutch will convince the Germans to turn a blind eye." What if it wasn't true? Maybe this was the moment when she was supposed to hand over the diamond ring ... But would it be useful? The Dutch officer continued to speak with the German officer, still indicating her: those seconds were endless, the anxiety now bursting in her chest, causing her to feel as if she could no longer breathe. Finally, she heard yelling: "Go on, go on, hurry up!" She staggered as she took up her bags and left the train: it was over, she was in Holland. She was finally free.

Further Reading

R.L. Sime, *Lise Meitner, A Life in Physics* (University of California Press, Berkeley, 1997)

Oppenheimer. An Explosive Plan

Dr. Strangelove

On April 28, 1954, in spite of an annoying breeze coming up from the Potomac River, in Washington DC, it was a beautiful spring morning. In a room on the second floor of the rather ugly building belonging to the Atomic Energy Commission, an exceptional witness was being heard. Besides the bench for the jury, consisting of three members, there was barely room for two more tables facing each other: in one sat J. Robert Oppenheimer, together with his defense lawyer Lloyd Garrison; in the other, there was Roger Robb, the general prosecutor, known as an old fox of the courtrooms. The fourth man in this tableau, the man sitting in the witness chair that morning, did not seem at all perturbed. On the contrary, he gave the impression of being perfectly at ease in his role: dressed elegantly in a gray double-suit and slicked back hair, he articulated his words with extreme naturalness, accompanying them with eloquent gestures of satisfaction and the constant movement of his eyebrows, two thick black bushes placed right in the middle of his forehead. It seemed like he wished for nothing more than to be there that day, to finally be able to speak his piece: it is a well-worn maxim that revenge is a dish best served cold, and he was there to confirm it. He also appeared to get along very well with the attorney general, almost as if they had agreed upon the questions and answers ahead of time. In fact, when Roger Robb asked him,

Do you or do you not believe that Dr. Oppenheimer is a security risk for the United States?

his reply was almost instantaneous:

In a great number of instances, I've seen Dr. Oppenheimer act in a way which, for me, was exceedingly hard to understand, and his actions frankly appeared to me confused and complicated. To this extent, I feel I would like to see the vital interests of this country in hands which I understand better and therefore trust more. In this very limited sense, I would like to express a feeling that I would feel personally more secure if public matters would rest in other hands.

© Springer Nature Switzerland AG 2020
G. Mussardo, *The ABC's of Science*,
https://doi.org/10.1007/978-3-030-55169-8_15

In hearing that, Oppenheimer slowly looked up, his eyes appearing even more intensely blue than usual. He looked impassively at the man who sat opposite him, but in his mind, he went back to the summer of 1942, when he had first met him at Berkeley, and later at Los Alamos. He thought about how he had invited him to join an exceptional group of nuclear physicists, including Enrico Fermi, Hans Bethe, John van Vleck, Felix Block and John von Neumann, to study the possibility of building a terrible bomb, a bomb that, for the first time, would unleash all of the enormous energy contained in the very small core of an atom. The witness looked back at him obliquely. The sarcastic light that crossed his eyes for a second revealed, more than ever, the anger that he had held inside for many years, too many. The man who sat at the witness stand was Edward Teller, the true Dr. Strangelove.

A Man, An Enigma

J. Robert Oppenheimer (the J standing for Julius, his father's name) has always been an enigma to many. In fact, how do you get along with someone of such a complex and sophisticated personality? Someone who could be extremely arrogant yet intimately insecure, robust and fragile at the same time? A scientist with fulminous, yet dispersive, intelligence, the main reason, after all, why, despite his numerous and brilliant theoretical results, he was never part of the Olympus of twentieth-century physics in which Einstein, Dirac, Schrödinger and Pauli were major performers. As Pauli once remarked, "Oppenheimer seemed to treat physics as a recreation and psychoanalysis as a vocation." It was as if he were always in a restless and uneasy search for his true identity. Tall, handsome, with incredibly blue eyes, possessed of a hieratic aura enriched by a thousand quotations from Sanskrit or French literature, Oppenheimer was incredibly charismatic. It was almost impossible to remain indifferent to his person. There were typically two discordant reactions: the first—numerically the majority—saw him as a man with exceptional intellectual abilities, amalgamated by incredible charm and a unique facility for persuasion, driven, moreover, by a great moral integrity. The second, instead, was diametrically opposed, pointing the finger at his arrogance and his annoyingly presumptuous attitude, in particular, at his theatricality. His detractors used to express the embarrassing feeling of being in the presence of an actor, able to play whatever role was necessary for the sake of being at the center of the scene, roles that could change course according to the circumstances and that did not preclude the betrayal of friends, colleagues or his own country …️ These were the severe comments of his most bitter enemies: few in number, but, unfortunately, vast in influence.

Oppenheimer (also called Oppie by his friends) was born in New York on April 22, 1904, into a rich family of Jewish origin, although they were not observant: his father was a rich textile merchant who emigrated from Germany, while his mother was a very sophisticated New York woman, a lover of art and literature. The Oppenheimers lived in Manhattan on Riverside Drive, an elegant residential neighborhood along the banks of the Hudson River, in a luxurious apartment embellished with three van

Goghs, as well as paintings by other famous artists. Indeed, money was never a problem for Oppie …

He studied at the School of Ethical Culture, where, in addition to mathematics, chemistry and physics, he also learned ancient Greek, European literature and art history, all of which remained at the center of his interest. He later enrolled at Harvard, graduating with a degree in chemistry in 1925, after which, at the suggestion of his mentor at that time, Percy W. Bridgman, a future Nobel Prize in Physics, he went to Europe, to the Cavendish Laboratory in Cambridge, to work with JJ Thompson, the physicist who had discovered the electron in 1897. Whether due to Thompson's old age, the rather boring research topic assigned to him (studying the absorption of electrons by thin metal films), or, more simply, because experimental physics was not his cup of tea, Oppenheimer decided to stop switching galvanometers on and off, calibrating vacuum tubes and counting the ticking of Geiger counters, turning his attention instead to theoretical physics, a field that was on the verge of a great revolution, promising to be full of exciting surprises. In 1926, he decided to move to Göttingen, a small and pretty German town, at the time, one of the neuralgic centers of quantum mechanics, that new theory that aimed to get to the bottom of all of the bizarre phenomena of matter that occur at an atomic scale and that had completely absorbed the attention of Europe's most brilliant physicists.

It was almost perfect timing: indeed, during 1926, the original theory of atomic phenomena proposed in 1913 by Niels Bohr witnessed an impressive series of new developments, starting from the idea of the wave function introduced by Erwin Schrödinger, and followed by the uncertainty relation discovered by Werner Heisenberg. The previous year, the first consistent formulation of the new theory had been put forward by Max Born, director of the Göttingen Institute of Physics, Pascual Jordan and Werner Heisenberg: it was based on matrix algebra, although it was later shown to be perfectly equivalent to the wave mechanics proposed by Schrödinger. Furthermore, it was once again Born who proved that the wave function proposed by Schrödinger had a very evocative interpretation in terms of probability density. In other words, those were the golden years of quantum mechanics, and Oppenheimer happened to be right at the core of that world. In Göttingen, he was able to work with Max Born, who was his Ph.D. supervisor, in Leiden, he worked with Paul Ehrenfest, and in Zurich, he worked with Wolfgang Pauli, a scientist strongly recommended to him by Ehrenfest "in order to acquire more discipline and more mastery." During his years in Göttingen, Oppenheimer, in collaboration with Born, wrote an important paper on the quantum properties of the molecules: besides soon becoming one of the key references of molecular physics, this paper also made him rather famous.

Box: The Main Stages of the Atomic Age

1932 James Chadwick discovers the neutron.

1934 The Joliot-Curies discover artificial radioactivity and the possibility of transmuting chemical elements when subjected to alpha particle bombardment.

1934 Enrico Fermi systematically studies how chemical substances give rise to neutron bombardment When he reaches uranium, he reports that he has found several radioactive fragments that he was not able to identity.

1938 Otto Hahn and Fritz Strassmann prove that what Fermi observed four years earlier was the fission of the uranium nucleus in two lighter nuclei, including barium.

1939 Lise Meitner, together with her nephew Otto Robert Frisch, describes the fission mechanism of the nucleus and the large amount of energy that is released in such a process. The question remains as to whether the number of neutrons released by fission is sufficient to start a chain process.

1939 Niels Bohr gives an account of these sensational developments at a congress held in Washington.

1939 Leo Szilard immediately confirms that the neutrons released in the reaction are sufficient to give rise to a chain reaction. The Joliot-Curies, who publish their results in an issue of Nature, confirm the same.

1939 First letter from Einstein (at the request of Leo Szilard) to the President of the United States for the purpose of drawing his attention to the danger that German scientists could develop an atomic bomb of devastating effects.

1941 The MAUD report, in which Rudolf Peirls and Otto Frisch give an accurate estimate of the critical mass of uranium, is sent to the United States.

1942 The Manhattan Project is officially set up under General Leslie R. Groves and J. Robert Oppenheimer.

1945 After the Trinity Test in New Mexico, two atomic bombs are dropped on Hiroshima and Nagasaki in Japan.

1949 The Soviets conduct the first test of their atomic bomb in the desert of Kazakhstan.

1949 Klaus Fuchs confesses that he committed espionage for the Soviets during his stay in Los Alamos.

1950 President Harry Truman orders a hydrogen bomb project to be launched.

1952 The first US hydrogen bomb is tested on an atoll of the Marshall Islands in the Pacific Ocean. Edward Teller follows the test in a seismograph room at Berkeley.

1953 The Soviet-built hydrogen bomb (thanks to Andrei Sakharov's supervision) is tested in the Kazakh desert polygon.

1954 J. Robert Oppenheimer loses access to atomic secrets after a Maccharist-style witch-hunt.

With such a curriculum, he did not have much trouble finding a job upon his return to the United States: he received offers from many universities, including Harvard, but he decided to go to California, attracted by the possibility of creating the first real school of theoretical physics in the United States from scratch. Hence, he accepted a professorship at the University of Berkeley, partially in view of the fact that, "Berkeley owns a large collection of French poets from the sixteenth and seventeenth centuries." The agreement was that he could also spend some time in Los Angeles, at Pasadena, teaching at Caltech.

A Dedicated Anti-communist

Edward Teller was born in the heart of the old Mitteleuropa, more precisely, in Budapest, on January 15, 1908, also into a family of Jewish origin: his father, a wealthy lawyer, was well integrated into the academic and governmental environment of the Austro-Hungarian Empire, whereas his mother was a talented pianist who spoke several languages fluently. The end of the First World War not only disintegrated the cheerful atmosphere of the Empire, it also brought Hungary to the brink of poverty and civil war. The communist regime that took power in 1919 under Bela Kun immediately became known for its brutality, leaving a long trail of murder, detainment and torture. The climate of terror of those months heavily marked the Teller family, who were exposed to the continuous threat of expropriation of all of their property and the violence of the soldiers bivouacking in their home. This traumatic experience left an indelible trace on Edward Teller, who thenceforth developed a visceral hatred for anything that was even remotely associated with communism. The fall of the Bela Kun regime was followed by even more tragic and catastrophic years for Hungary, with the seizure of power by an old chap of the Magyar aristocracy, the Admiral Miklos Horthy, who continued with the murders and tortures of his political adversaries, setting up a ferocious fascist government, the first in twentieth century Europe. "During my first eleven years," wrote Edward Teller, "I knew war, patriotism, communism, revolution, anti-Semitism, fascism and peace. I just wish the peace had been much longer."

Since he was a child, he had shown an amazing intelligence and great versatility in mathematics and other scientific subjects. Due to the stifling political climate in Hungary and the restriction of university access to Jews imposed by the authorities, Teller decided to move to Karlsruhe, in Germany, to study chemistry. But, in 1928, he became fascinated by what was happening in physics, and he therefore left behind both Karlsruhe and chemistry to move to Munich, in order to work with Arnold Sommerfeld, one of the pioneers of quantum mechanics that was ascending in those years. However, his interaction with Sommerfeld was not ideal: Teller's exuberant character, combined with his aggressiveness, did not match well with the composure and detachment of such an old-school German professor as Sommerfeld.

As chance would have it, a sharp turn of events occurred: in Munich, in an attempt to catch a tram, Teller slipped on the tracks and the vehicle severed his foot. The surgeon who operated on him was very clever, implanting a prosthesis on what remained of his bone, but before he could stand again, Teller had to spend several months in the hospital. During his convalescence, he was visited several times by Hans Bethe, whom he had met in Munich: despite his young age, Bethe already had a reputation as a great scientist, not so much for his genius as for his tireless and infallible calculation skills. "Bethe's brain," said Teller "is like a diesel engine, slow but relentless." Once healed, in October 1928, Teller decided to leave Munich behind and to go to Leipzig instead, to the physics institute directed by Werner Heisenberg. "I came into contact with the highest form of knowledge, the one I had always dreamed of when I was a child," he commented later.

In Munich, he realized that he had joined an exclusive club of young physicists, at the head of which was Heisenberg, the rising star of theoretical physics. In addition to the Soviet physicist Lev Landau, he also had the opportunity to meet the Swiss scientist Felix Bloch and the American John van Vleck, whom he would meet again years later in Los Alamos. He also met Carl von Weizsäcker, a Prussian aristocrat destined to be, in the years that followed, one of the main characters of the atomic project carried out by Nazi Germany. In 1932, Teller spent the entire summer in Rome with Enrico Fermi, an experience that impressed him enormously for the intellectual clarity and the simple manners of that great Italian scholar. With Hitler's seizure of power in 1933, Teller realized that it was time to leave Germany and get as far away as possible. The opportunity came in 1934, when, thanks to the colourful character responding to the name of George Gamow (who, in the meantime, had escaped to the United States from Stalin's Soviet Union), Teller was offered a chair of physics at George Washington University in Washington DC. On the ocean liner that he and his wife took to go to the United States, Teller met back up Hans Bethe, who was also escaping from Nazi Germany.

Professor at Berkeley

In the '30s, California was like paradise on earth: the mountains of the Sierra Nevada, the long beaches with their high cliffs along Highway 101, the Mojave desert, the sweetness of the Napa Valley landscape and the breathtaking landscapes of the great national parks, all of this made that corner of the planet an extraordinary place. There were also splendid urban landscapes: Los Angeles, as portrayed by Raymond Chandler, with its eucalyptus trees, the hills around the nascent Hollywood Studios, Mulholland Drive and the Beverly Hills villas, were matched by San Francisco, with its morning fogs, colourful Victorian houses, long marina premises, and the shrieks of sea lions lying in the sun at Pier 39. Needless to say, Oppenheimer immediately fell in love with those places. He spent most of his time in Berkeley, even though he frequently went to Caltech in Pasadena. Thanks to him, the Berkeley campus became, in a few years, one of the most vibrant physics centres in the United States. Together with his extraordinary teaching qualities and his passion for theoretical physics, Oppie also had a genuine interest in the nuclear physics experiments carried out at Berkeley by Ernest Lawrence, of whom he became a good friend, despite their obviously divergent backgrounds: while Oppie was a well-educated and refined theoretical physicist, a rich Jew from the East Coast, Lawrence was a very skilled experimental physicist who came from the deep belly of America, South Dakota, and who had to pay for his studies by working as a door-to-door salesman. The students adored Oppie, to the point of imitating his unmistakable walk or his way of lighting his pipe: they also copied his characteristic expression "ja-ja,..., ja-ja," pronounced with a slight German accent, that punctuated his conversations. Oppenheimer met Jean Tatlock in the mid- '30s and, if he had never been interested in politics until then, thanks to this young woman, who was studying psychology at Berkeley, he soon came into contact with left-wing political groups, circles made up of professors and intellectuals who loved to write about the Great Depression, international politics and the fascist regimes that were taking power in Europe, spreading the word in various newspapers that circulated more or less clandestinely. He also approached Trade Union groups that were fighting for the rights of the teachers and wrote several articles on their behalf, even though he never signed them. In 1937, he inherited a large sum of money upon his father's death and, along with other initiatives in support of the radical circles in California, he was also happy to generously finance all of those who fought alongside the Republicans in the Spanish Civil War. His brother, Frank Oppenheimer, and his wife Jackie became members of the communist party in those years, while Jean Tatlock had long since been a member. However, Robert and Jean's love story ended badly, dissolving in the first months of 1939. In August 1939, at a party held in Pasadena, Oppenheimer met Kitty Puening, the newlywed spouse of Steward Harrison, an English doctor living in the United States. Kitty had previously been married to Joe Dallet, a fervent American communist who died in Zaragoza in 1937 during the Spanish Civil War. That meeting in Pasadena was a real electric shock for both of them: after a few months, Kitty divorced her husband and married Robert Oppenheimer. The haste of that love story left many people scandalized and others stunned, the latter, for some, because Oppie had actually proved to be human.

The Atomic Era

The seminar held by Niels Bohr in Washington on January 26, 1939, caused a great sensation among the American nuclear physicists: the great Danish scientist announced that two German physicists, Otto Hahn and Fritz Strassmann, had succeeded in splitting the nuclei of uranium into smaller fragments, releasing, in the process, an enormous amount of energy. Among those present at the seminar, many rushed to their laboratories to confirm what they had just heard. The atomic era had truly begun. With the outbreak of war and the growing terror that physicists left in Germany, under the brilliant guide of Werner Heisenberg, could be the first to build an atomic bomb of appalling power, in June 1942, the United States finally set up the Manhattan Project. But to achieve this, some initial steps were required: the famous letter from Albert Einstein to the president of the United States, then the incessant efforts in political circles by Leo Szilard, and finally the reading in Washington of the British MAUD report, in which, for the first time, an accurate estimate of the critical amount of uranium needed to build an atomic bomb was put down in writing.

While Enrico Fermi created the first atomic pile using a controlled nuclear reaction in December 1942, in the summer of that same year, J. Robert Oppenheimer had assembled a small group of brilliant young nuclear physicists at Berkeley to study the feasibility of this new bomb. The meetings took place in a room on the top floor of Le Conte Hall, the building where Oppenheimer had an office, and all precautions were taken to ensure the secrecy of what was being discussed in that room. Among those present was Edward Teller, who tried to bring everyone's attention to the possibility of building a bomb with virtually unlimited power, thanks to a reverse process of the fission of uranium atoms, i.e., by using the fusion of hydrogen atoms into helium with the corresponding release of a large amount of energy: in a nutshell, it is the same nuclear reaction that takes place in the Sun and all of the other stars, making them shine. The discussions among the members of the group became increasingly animated, and the atmosphere became more and more tense: finally, Oppenheimer succeeded in putting Teller in the corner, pointing out that the explosion of a bomb like the one conceived by Teller could be so catastrophic as to destroy Earth's entire atmosphere! It was the first time that Teller had to swallow a bitter pill.

The second time was at Los Alamos. General Leslie R. Groves, who was the head of the Manhattan project, decided to employ Oppenheimer as the scientific director of that titanic enterprise, despite the FBI's reports on his "liberal" background. Thanks to his extraordinary persuasive power, Oppenheimer was quickly able to bring to Los Alamos, a remote place in New Mexico, "the largest egghead collection," a group that included Enrico Fermi, Hans Bethe, Robert Wilson, Emilio Segrè, Ernest Lawrence, Richard Feynman, John von Neumann and Edward Teller. Teller had arrived in Los Alamos eager to work on "his" bomb, the SUPER, as it was called for its terrific power. But, once again, Oppenheimer had decided otherwise: instead, Teller would have to work under the supervision of Hans Bethe, who was appointed director of the theoretical division of the laboratory. Teller's reaction was vehement: with his typical aggressiveness, he used fiery words to turn all of his anger on Oppenheimer. He said that he had come to Los Alamos for his bomb, not to work on problems that any student could easily solve, and, much worse, under the guidance of someone else, in particular, Hans Bethe, whom he considered inferior to himself.

Faced with that endless flood of words, Oppenheimer preferred to be accommodating: instead of sending him away from Los Alamos, he decided to isolate him from the rest of the group and entrust him with the task of carrying out the SUPER studies on his own. The whole laboratory would instead work on the construction of an atomic bomb based on the fission of the uranium atoms. That bomb was finally built and tested on July 16, 1945, in the Alamogordo desert, in the so-called "Trinity Test." The show was impressive, and Oppenheimer described that moment with a verse from a Sanskrit poem: "Now I have become Death, the destroyer of worlds." The first atomic bomb was dropped on Hiroshima on August 6, 1945, the second on Nagasaki on August 9, 1945.

A Sham Trial

The news that arrived later from Japan on the devastating effects of the bomb on civilians threw many physicists into despair, Oppenheimer among them. In the years after the war, in the meetings of the various atomic energy committees that he was chairing, Oppie tried in every way possible to promote international control of atomic energy and a reduction in nuclear weapons. And this became his third arena of confrontation with Edward Teller, who was spending more and more time lobbying the military and Washington circles in support of realization of the SUPER. The turning point was the year 1949, when public opinion was shaken by two bits of breaking news, one more shocking than the other: the first was that the Soviets also had the atomic bomb and that they had tested it in the deserts of Kazakhstan; the other bit of news was that Klaus Fuchs confessed that he had passed atomic secrets to the Russians when he was in Los Alamos during the war. Both items threw the United States into a state of panic, and the hysteria that followed gave birth to one of the darkest and most paranoid periods of that country: McCarthyism. Anyone remotely suspected of having communist sympathies was placed under FBI surveillance, reported upon, interrogated, jailed, or even sent to the electric chair, as was the case in 1953 with the couple Julius and Ethel Rosenberg. In his tireless advisory role to the President of the United States on atomic issues, Oppenheimer had made bitter enemies over the years, such as J. Edgar Hoover, the powerful head of the FBI, the senior officers of the Navy Force, eager to enter into the management of nuclear weapons, and Lewis Straus, the new head of the Atomic Energy Commission that Oppenheimer had ridiculed in a government hearing on the use of nuclear isotopes. It was easy for all of them to find someone to dig up the old FBI reports on Oppenheimer's "liberal" sympathies and put him in the corner, giving him a drastic ultimatum: either relinquish his influence on US nuclear policy once and for all, implicitly admitting his communist sympathies and perhaps some further involvement with the Soviets; or face a "hearing" to determine whether he would be allowed to continue to have access to all of the nation's atomic secrets. For all that he had done for the United States, Oppenheimer believed that the first choice was an admission of guilt, and therefore he decided on the hearing. The trial that followed was a farce, marvelously orchestrated by Lewis Straus with the help of the ruthless forensic art of Roger Robb, the general prosecutor. The scientists who were called to testify—Fermi, Rabi, Bethe, and many others—all had words of praise for Oppie and for what he had done for the United States. The only contrary voice was Edward Teller's, but it was the decisive one in bringing about the final verdict of condemnation.

Further Reading

R. Monk, *Robert Oppenheimer* (Knopf Doubleday Publishing Group A Life Inside the Center, 2014)

Pauli. An Odd Couple

That Wolfgang Pauli was one of the greatest theoretical physicists of the twentieth century is a well-known fact: his discoveries—in particular, the famous exclusion principle (which earned him the Nobel Prize in 1945) and the hypothesis of the existence of the neutrino—are the basis of our current understanding of matter and among the most important developments in the field of elementary particles, astrophysics and cosmology. He was a decidedly colorful character, so much so that Oppenheimer once said that Pauli was the only person he knew who was equal to his own caricature. He was renowned not only for his great scientific acumen, but also for his sarcastic and relentlessly critical spirit, the source of innumerable anecdotes, among other things. His caustic remarks are famous: "So young and already so inconclusive," addressed to an unfortunate theoretical physicist, as well as "He is more interesting when he is drunk than when he is sober"; of Rudolf Peierls, a physicist of German origin and his colleague at ETH Zurich for some time, he once said, "He speaks so fast that before you've understood what he said, he's already told you why it's wrong." Mockery and jest were his favorite weapons. The pungent sentences and the fierce criticism were, in fact, the result of the family environment in which he grew up, in which everything, absolutely everything, was discussed and criticized. "Working with Pauli was absolutely fabulous. You could ask him anything, there was no danger that a particular question would be stupid; for Pauli, all questions were stupid," remarked Victor Weisskopf, one of his closest collaborators. Werner Heisenberg never published a work without first having asked Pauli for his opinion, and many others did the same: for this, in addition to the nickname "the Punishment of God" (with which he signed his letters on a number of occasions), Pauli was often called "the conscience of physics."

The "Pauli effect" was also well known: many took it for granted that a disaster would occur in one's experimental devices if Pauli's presence was sensed nearby. Peierls gave a description of this mysterious effect that escaped the logic of any physical law: "It was a sort of curse that was exercised on people or objects nearby, especially in physics laboratories, causing accidents of every kind. When he arrived in a laboratory, the machines stopped working, the glass test tubes cracked, a leak

© Springer Nature Switzerland AG 2020
G. Mussardo, *The ABC's of Science*,
https://doi.org/10.1007/978-3-030-55169-8_16

was created in the vacuum system, but none of these accidents directly damaged or disturbed Pauli." The experimental physicist Otto Stern even categorically forbade him from approaching his laboratory. Pauli's arrival in Princeton in 1950 was announced by a fire, for no apparent reason, in a newly installed, and expensive, cyclotron. But the most striking confirmation of the Pauli effect took place when the instrumentation in the laboratory of James Franck of the Physics Institute of the University of Göttingen suddenly burst, shattering oscilloscopes, counters and ammeters for no obvious reason. "If I didn't know that Pauli is a thousand miles away, I would say he's in the middle of it," Franck ventured jokingly. Only later did he learn that the disaster had occurred at the very instant when the train carrying Pauli from Zurich to Copenhagen made a five-minute stop at the Göttingen railway station. The evil Pauli had struck again.

Black and White

Wolfgang Ernst Pauli was born in Vienna on April 25, 1900, the only child of a respectable Jewish family originally from Prague. His father was a famous chemist who had built an international reputation in the protein field. He changed the surname Pascheles to Pauli in 1898, and the following year, he converted to Catholicism, a rather common choice for a Jewish-born academician in the Austro-Hungarian Empire: when, however, Pauli discovered this fact many years later, the effect was traumatic, and it was a topic that he simply didn't care to discuss. His middle name, Ernst, was in honor of Ernst Mach, his baptismal godfather ("a baptism against metaphysics," Pauli liked to joke). Since he was a boy, Pauli had distinguished himself for his ability in physics and mathematics. He also had a penchant for the history of antiquity, but when he had had enough of the Greek and Latin classics, he would immerse himself in reading Einstein's articles, spending his afternoons meditating on general relativity. By the end of high school, he had written three articles on this new and fascinating theory of gravitation, in which masses curve the space–time that, in turn, curves their trajectories: Einstein had found a completely different way of understanding gravity, of thinking about the arena of the physical world, about the role played by energy and geometry. In simple terms, he had written the most beautiful theory ever. Pauli was fascinated by it. He therefore decided that he wanted to continue studying physics at all costs, but it seemed to him that Vienna offered no stimuli. Better then to go to Monaco, to see Arnold Sommerfeld, whom he had met at the age of twelve at one of his conferences in Vienna.

Arnold Sommerfeld was a renowned master, the perfect embodiment of the German university professor, always well dressed, with his penetrating gaze, his face dominated by a beautiful martial mustache, perfectly in order on each side of his nose. Over the years, he served as the undisputed mentor of entire generations of theoretical physicists, seven of whom were later awarded the Nobel Prize: he loved to involve his most brilliant students in his research, putting them in contact with the

most famous scientists, and then remaining in contact with them himself by letter. The turning point of Sommerfeld's career was 1913, when, at the height of his forties, he learned of Niels Bohr's atomic model. He encouraged all of his students to go deeper into that subject; he, for his part, contributed significant improvements and immersed himself in the systematic and very detailed analysis of atomic spectra. His results formed the guidelines for a monumental text with which all of the quantum physicists of the time were trained: *Atombau und Spektrallinien*, the so-called Sommerfeld Bible. When Pauli arrived in Munich in 1918, the city was in the utmost chaos, Germany was one step away from capitulation in the war that had caused almost 40 million deaths in Europe and had shattered all geopolitical balances. The population was exhausted from those years of conflict and hungry from the months-long absence of crops from the countryside: the socialists had an easy time organizing an imposing street demonstration, asking that the king abdicate in order to give rise to a republic along the lines of the Bolshevik republic that had taken place in Russia. The king actually fled, and a socialist republic took hold in Munich, but it lasted less than a year and was forcibly suppressed by the federal government of Berlin.

The echo of these events obviously also reached the corridors of the university, and yet the physicists' heads were all turned toward the amazing discoveries of Einstein, Planck, Bohr ... When Pauli met Sommerfeld, the latter understood that he had very little to teach him: he immediately entrusted him with the task of drafting a review article on general relativity, an article that was later published in the Encyclopedia of Mathematical Sciences and that attracted the attention and praise of the greatest experts in the field, including Einstein, Weyl and Eddington. Pauli spent years of intense study in Munich, in the company of another young scientist destined to leave a profound mark on physics, Werner Heisenberg. The two immediately struck up a strong friendship, despite their irreconcilable lifestyles. If Heisenberg was as clear as water, with his country boy demeanor, his blond hair cut like a brush and his eyes clear and bright, Pauli was, instead, as dark as the night: in Berlin, he had since long discovered the bohemian district, where everything seemed lawful, and he loved to spend his time there, drinking, making new acquaintances, attending the entraîneuses, listening to music, and then returning home in the middle of the night and starting to work until the early hours of the morning ...

In Munich, he was full absorbed by the problem of generalizing Bohr's theory of the hydrogen atom to more complicated systems, such as the helium atom, and understanding how those strange atomic laws worked. His dazzling career had just started. Pauli was one of the most itinerant of quantum physicists: he soon moved from Munich to Gottingen, in the group led by Max Born, and then to Copenhagen, so as to be in direct contact with Niels Bohr. The environment there was a pleasant surprise. If, in Munich and Göttingen, you learned to do the calculations—said the physicists of the time—in Copenhagen, you learned to think. Later, he moved to Hamburg, and finally to Zurich, where, in 1928, at the age of twenty-eight, he became professor of physics at the Federal Polytechnic (ETH), the same university where Einstein had studied years before. Except for a five-year period, from 1940 to 1945, spent at Princeton at the Institute for Advanced Study, he remained in Zurich for the rest of his life, dying there on December 15, 1958, in hospital room 137, curiously, a number that regulates the quantum processes in which electrons and photons are involved, a number that was the obsession of his whole life.

Two Brilliant Intuitions

Needless to say, wherever he went, he left his mark. He would announce himself with his loud and slightly sardonic laugh, and he was always brimming with new ideas that he would proclaim as he walked incessantly back and forth, his plump body swaying slightly in this dance. If, in Munich, the partnership between Sommerfeld and Pauli had been a happy one, the one that he established with Bohr in Copenhagen was even more stimulating, because he had started to touch on the mysteries of the new quantum theory. In the 1920s, the state-of-the-art in this field was somewhat confused: Bohr's famous article from 1913 had opened revolutionary scenarios for the atomic world, described in terms of discrete energies and quantum leaps, but many unresolved and varied questions remained open, and even paradoxical. The attention of the scientists had been concentrated, in particular, on the puzzle of "quantum numbers": above all, they wanted to understand how many quantum numbers were needed to characterize the electronic states of the atom, in order to be able to predict the physical properties of the chemical elements, as well as the sequence of the spectral lines. However, it was proving impossible to find the key to the story: if only one quantum number seemed to be sufficient for the hydrogen atom, it turned out that at least two were needed for the helium atom, while three or even four quantum numbers were needed to describe the heavier atoms.

Bohr had also proposed that Pauli work on a very strange phenomenon, discovered a few years earlier by Pieter Zeeman, a young Dutch researcher from Leiden: Zeeman had noticed that, by placing atoms in a magnetic field, each of their spectral lines separated into a multiplet of lines. It was the famous "anomalous Zeeman effect". Bohr wanted to clarify it, and Pauli seemed the right person to understand the matter thoroughly. It was in that circumstance that, for the first time, Pauli came across the number 137. Sommerfeld had introduced it to describe the fine structure of the energy levels of the atom; more precisely, he had introduced its inverse, 1/137, a quite small number built using three of the most fundamental constants of nature: the electric charge of the electrode, the speed of light and the Planck constant, all of them at the basis of the interaction of light with matter.

It was understood that the two problems, that of quantum numbers and that of the Zeeman effect, were somehow deeply connected. Heisenberg too, urged by Sommerfeld, was working on it, and, very quickly, he had produced a strange theory that predicted that the angular momentum of the nucleus and electron had to be quantized in terms of semi-integer quantities, such as 1/2, 3/2, etc. Things seemed to be working, but it was not clear why. Pauli was unable to keep his mouth shut, even on that occasion, when his best friend was involved. He publicly mocked him by saying that, in Germany, atomic physicists were divided into two categories: the first used integers and, if the results did not materialize, they would redo the computations by replacing them with semi-integers; the second, on the other hand, first used semi-integers to make their calculations and, if the results did not materialize, they did it again using integers this time. Finally, it was Pauli's turn to untangle that mysterious skein, and he did, thanks to two brilliant intuitions. At the time, in 1924, he was in Hamburg, immersed, as usual, in his messy life, as a regular customer at bars, cabaret shows, and red light district brothels, stinking of beer and alcohol and sometimes becoming involved in fights and scuffles. Yet, playing with impunity within that shadow realm, Pauli suddenly saw the fog disappear and the light shine through, clear, crystalline, clarifying the problem that had haunted him since he had heard it for the first time.

As for the first brilliant intuition, he hypothesized that the electron, considered, until then, as a point characterized only by mass and electric charge, was instead like a small top, and that it therefore had an internal degree of freedom, also quantized, associated with its rotation around an axis. A different spin orientation—as this new degree of freedom is referred to—immediately provided the explanation of the mysterious ghostly lines of the Zeeman effect. Furthermore, regarding the problem of quantum numbers, Pauli formulated the famous exclusion principle that bears his name: two electrons can never share their quantum numbers. This principle immediately accounted for the progressive size of the atoms, gradually rising with the atomic number, and gave reason for all of the regularities observed experimentally concerning the filling of the atomic shells: in a few words, this principle provided the key to understanding the periodic table of the elements! Bohr's reaction had been very encouraging; hearing it, he said it was "a true madness," which, in his jargon, meant that he was deeply happy and believed it to be damn correct.

A Trunk Full of People

To appreciate an extraordinary character such as Pauli (his name even pops up in the pages of the diaries of Elias Canetti, who described him as a regular visitor to the literary and philosophical conferences that animated the cultural life of Zurich in the 1930s), the most valuable source undoubtedly turns out to be his collection of letters, most of which are kept in the Pauli Archive at CERN, in Geneva. This is a wealth of correspondence with many scientists and illustrious personalities of the time, a real gold mine that reveals more than one surprise: in fact, the more Pauli appears to be sober, austere, and extremely controlled in his scientific articles, the more compulsive he comes across in his letters, drawn up in a colloquial and often highly speculative style. Looking at the correspondence, one has the impression of being in front of a great obsessive scribbler, and one cannot help but think of Fernando Pessoa and his famous trunk full of people … a comparison that is by no means inappropriate, as we'll see.

The gem of this gold mine is the exchange he carried on for thirty years with Carl Jung, the controversial founder of analytical psychoanalysis. This correspondence offers an exceptional document for understanding both quantum mechanics and psychoanalysis, two fields of investigation apparently without any point of contact, not even of a methodological nature. Psychoanalysis is concerned with describing psychic processes through psychic means: the psyche is therefore simultaneously the object and the subject of study, the means and the end, and there is no room for conducting any investigation "from the outside". Physics, on the other hand, tries to investigate material reality in the most objective way possible, and therefore relies on measurements and experiments, as well as on the powerful tools of mathematics. This objectivity—it must, however, be recognized—is illusory; in fact, physics uses the extraordinary capacities of the mind both to conceive new experiments and to proceed with their elaboration. If it is true that Physics is based on the impersonal data of the measurements, the question of their interpretation still strongly depends on the nature of the human mind. Put simply, the vision of our Universe is deeply and inevitably of an anthropomorphic nature.

In their ten-year dialogue, which also resulted in the publication of the collaborative book *The Interpretation of Nature and the Psyche*, Pauli and Jung show their utmost interest in crossing the columns of Hercules of their respective fields of investigation and in seeking, with humility but also with the necessary audacity, the link between observable and still unknown reality. In this regard, Jung's omnivorous attitude is well known: he never hesitated to incorporate any element into his research that could be useful for understanding human behavior, be it anthropology, mysticism, cabalism or spiritualism. So, why not physics? Pauli, for his part, was fascinated by the mental mechanisms that lead to new theories, that is, he was interested not only in the philosophical aspects of the theory of relativity and quantum mechanics, but also in their psychological implications. If Pauli's passion for the analysis of physical processes can be ascribed to the clarity with which he sought their final formulation, he was, however, deeply fascinated by the problem of establishing contact with the field of creativity and spiritual processes, an area in which—we know—there are no rational formulas …

The Hardened Scientific Rationalist and the Alienist

The above could leave the impression of a purely academic exchange between two of the greatest minds of the twentieth century. As a matter of fact, it was not: it is worth mentioning the circumstances of their first meeting, which took place in 1930, when Pauli, in a state of despair, contacted Jung to try to cope with a devastating period in his life. Both his mother's suicide and the subsequent failure of his very short marriage to a dancer had pushed him down the path of alcoholism and the dangerous nightlife of Zurich's bars and night clubs. All of these events had also sharpened his irritability to the maximum point, leading him to repeated unpleasant episodes, including physical disagreements, during those fun nights in the city's clubs. Jung remembered their first meeting with these words: "A hardened scientific rationalist had to consult an alienist, being close to losing his mind, because he had suddenly been taken by storm by the most incredible dreams and visions … When the aforementioned rationalist came to consult me for the first time, he was in such a state of panic that not only he but I myself felt the wind of hospitalization for the mentally ill blow."

Jung believed that it was better to entrust Pauli to the care of Erna Rosenbaum, a young and brilliant pupil of Austrian origin, citing the excuse that he had no time to deal with Pauli's problems himself, although this redirection had been done quite on purpose. At first, Pauli did not take it well, so much so that he wrote to Rosenbaum: "I had turned to Jung because of certain neurotic phenomena related to the fact that it is easier for me to succeed in academia than with women. But since it's the opposite for Jung, he seemed to me the right person to turn to. And here instead he told me to turn to you, a woman!" In spite of Pauli's initial diffidence, however, Rosenbaum immediately established a good relationship with him, to the point that she convinced him to transcribe the content of his dreams. In the period in which he was in analysis, Pauli transcribed and tried to interpret hundreds of dreams himself, eventually passing all of this documentation along to Jung, who could not believe his great fortune in being able to draw from such a precious resource for his research on the psyche and archetypes. Most of these dreams were, in fact, reported by Jung during his Eranos lessons, although Pauli's identity remained hidden behind elusive phrases such as "an important scientist," "a man of great intellectual ingenuity," etc. Pauli's unconscious was immense; apparently devoid of an exterior or an interior, it was turned towards the most distant past, from which sprang its Jewish roots, and its gloomy present.

Pauli felt that he was the custodian of a secret wisdom and an uneasiness of the soul that macerated him. In his dreams, his demons manifested themselves in snake figures, which, biting their tails, regenerated themselves; in figures of oriental women, who, dancing, accompanied him along a dark corridor that led to a large classroom crowded with people; in the device of a pendulum, which, without friction, moved incessantly, as if in perpetual motion; in the figure of the mother who poured water from one basin to another; in the nightmare of a large square, where, in its four corners, ferocious dogs sat biting anyone who passed by; in the scene in which Pauli himself read an ancient book on the Inquisition trials with Giordano Bruno and Galileo Galilei; and in his greatest vision ever, that of the world clock. The path of analysis was only partially effective: a few years later, Pauli happily remarried, regaining his psychological balance, but he never stopped drinking.

Like Two Communicating Vessels

The link between Pauli and Jung, however, continued, strengthening even after the end of the therapeutic period, resulting in a dialogue between "equals" that is unprecedented in the history of science. How can we not be fascinated in observing their progressive reciprocal influence, the game of communicating vessels that the course of time established between them?

Since the natural world is an aggregate of particles, it is of the utmost importance to find out where and how, for example, the photon (just to mention one example) can help us acquire a definite knowledge of reality through an exchange of energy processes. Light is simultaneously a wave and a particle, and this forces us to abandon a causal description of nature in favor of its probabilistic vision.

Is this Pauli speaking? No, it's Jung!

The division and reduction of symmetry is thus the real heart of the question. The first is an ancient attribution of the devil, oh, if only the two divine rivals—Christ and Satan—could notice that they have grown so symmetrically …

Jung? Wrong, Pauli!

This was their choreography: both of them were mobile, unpredictable, multi-faceted, changing their minds and their words. Robert Oppenheimer, who met Pauli in those years, reported that he treated physics as a form of leisure and psychoanalysis as a vocation. It is thanks to Pauli, for example, that Jung refined and formulated his concept of "synchronism." "What is decisive for me," wrote Pauli, "is that I dream of physics like Mr. Jung and others think of physics. Every time I talked to Mr. Jung about the synchronic phenomenon and the like, there has always been some spiritual fertilization."

Some even went as far as to see a clear alchemical influence in the idea of the exclusion principle, formulated by Pauli in 1925 (a thesis that would need to be demonstrated, if only for the obvious chronological discrepancy with the date of their meeting). However, it remains true that all of Pauli's scientific work was always centered on the problems related to the laws of symmetry: symmetry under a reflection in a mirror or changing the arrow of time, or even more abstract symmetries, such as those that link particles to their antiparticles. For Pauli, all of this was the epitome of beauty and truth of natural laws. And if, for alchemists of the Middle Ages, their interest was in sulfur, mercury and the philosopher's stone, modern physicists have learned to deal with the infinite transmutations of elementary particles. For Pauli, however, the truth was not to be sought in this variegated taxonomy of pieces of matter, but in the manifestation of something deeper: an Unus Mundus, to put it in Jung's words, an archetype, or the realm of symmetries and perfect order, where mind and matter originate.

Further Reading

C.A. Meyer, *Atom and Archetype: The Pauli/Jung Letters, 1932–1958* (Princeton University Press, Princeton, 2001)

Quantum. Erwin's Version

1925 had been a year without any noteworthy events, and as it had passed, so did it smoothly proceed towards its end. The Swiss countryside was covered with snow, a thick and very white blanket lying on the tops and along the slopes of the mountains like a layer of marzipan: if you looked at the woods, with a little luck, you could spot a few bears; turning your attention downstream towards the town in the distance, you could see the smoke from the chimneys of pretty chalets, sledges pulled by horses and the usual Christmas bustle. Arosa, the pearl of the canton of Grisons, never disappointed its guests: nestled between the peaks of the Weisshorn mountain range, it offered to visitors the pleasure of its breathtaking landscapes, long ski slopes, mountain-climbing, dinners based on fondue and mulled wine and the right atmosphere for romantic evenings in front of the fireplace.

A few days before Christmas, a strange couple showed up at the door of the hotel that belonged to a certain Dr. Herwig: she was little more than a teenager, but was on the verge of a femininity that promised to be bursting, an angelic face embellished with two large blue eyes and a series of curls that fell down onto her cheeks; he, a man of about forty years of age, with strange zouava trousers and a raw wool cape that covered his shoulders, his hair spiky, his face dominated by a spacious forehead and an important nose, all embellished by a pair of round black-rimmed glasses. Dr. Herwig warmly welcomed him; the man had been his guest before. Of course, the previous time, he had been with his wife, but there was no need to quibble or go into personal matters: what if he had divorced in the meantime? Or what if this young lady was just his granddaughter? In any case, it wasn't Herwig's business. After placing the suitcases on the floor and hurrying to the reception, the man asked the hotel keeper in a low voice if it was possible to avoid the hassle of registering the lady who accompanied him. "But of course, no problem, don't worry, please sign here and take the room key in the meantime, the room I have reserved for you is the most beautiful and quiet one, as you requested". Satisfied with the answer, the man turned towards the girl with a smile, then turned back to the entrance desk, took the pen, dipped it in ink and, in a diminutive signature, wrote *Erwin Schrödinger* on the register. The following three weeks were marked by skiing activity in the

© Springer Nature Switzerland AG 2020
G. Mussardo, *The ABC's of Science*,
https://doi.org/10.1007/978-3-030-55169-8_17

morning, intimate dinners in the evening, and passionate embraces at night, but, as his friend Hermann Weyl, who knew him very well, said, "in that outbreak of late love, Schrödinger completed that transcendental work that was going to change the course of physics forever."

The Theater of the World

Erwin Schrödinger was born in Vienna on August 12, 1887, the only son of Rudolf Schrödinger and Georgine Bauer. His father was a passionate lover of Italian paintings, as well as a great collector of oriental ceramics and a scholar of the evolutionary history of animal species. With regret, however, he had to sacrifice these interests in order to take care of the small family business in the linoleum trade, an activity that guaranteed a comfortable life for the family, at least until the end of the Habsburg Empire. The maternal side of the family was of Anglo-Saxon descent, and little Erwin, thanks to his aunt Minnie and her stories from the Bible, learned English before he even started to write in German. Furthermore, his maternal grandfather was a famous chemistry professor at the University of Vienna, well integrated into the high society of the capital: at one point, he had had the opportunity to buy an entire five-story building in an exclusive neighbourhood of the city, overlooking the steeples of St. Stephen's Cathedral, and it was on the top floor of that building that Erwin grew up.

As a child, he was always surrounded by the love and affection of the many women in his life: his mother, his aunts Minnie and Rhoda, not to mention the crowd of nannies and servants who milled about the house. The female universe thus became an essential spring of his life; the fascination for that world never left him, often serving as an extremely powerful creative stimulus, but also resulting in an exuberant sentimentalism that he manifested over the years in the courting of his many lovers. Each person, Gabriel García Márquez used to say, has a public aspect, a private one and a secret one: respecting the secret side, and therefore dealing only with the private one, Schrödinger's life unfolds before our eyes like the story of a daring seducer, a Don Juan always attracted to young and beautiful women, in a dizzying whirlwind of passions and loves that was nothing but the insensate and overwhelming force of life. One of his first conquests was the sister of his best friend from high school, Lotte, a girl with whom he flirted for an entire summer. Years later, he lost his head for a girl from the Viennese aristocracy, Felicie Krauss: Erwin told her that he was ready to abandon his academic career and everything else just to marry her, but the proposition came to an end due to the firm opposition of his father—who did not want his son to repeat the mistake that he had made years earlier—and an equally decisive refusal by the girl's mother, relentlessly reluctant to accept any ordinary bourgeois into the house, without any court lineage. The disappointment burned through him, and it was then that Erwin became convinced that true love perhaps should be sought through a conventional marriage, a belief that he scrupulously followed for the rest of his life.

Since he was a boy, he had showed an innate talent for learning; his classmates always referred to him on every matter of mathematics and physics, his favourite subjects, together with philosophy. He enrolled in the course of Physics at the University of Vienna in 1906. It was the same year in which Ludwig Boltzmann had decided to end his life on the shores of the Adriatic, but, in spite of his disappearance, the great scientist continued to exercise a tangible influence within the university and on all those who approached science: for example, Fritz Hasenöhrl and Franz Exter, the professors who followed Schrödinger during his graduation thesis, came from Boltzmann's school and were also enthusiasts of thermodynamics and statistical physics, a passion that they passed on to their pupil.

"In physics, there are not only atoms, in science, there is not only physics, in life, there is not only science," Schrödinger used to say; in accordance with his saying, from an early age, his interests widened dramatically into many fields, including the most distant horizons of knowledge, the deepest questions of philosophy and the most lovable pleasures of life, including art, literature and a taste for adventure, enhanced with a certain rebellious streak. He was obviously not the only free or transgressive spirit around; that was, after all, the Viennese spirit of the time. Karl Krauss did not hesitate to define those years as "the laboratory of the end of the world": Gustav Klimt and Egon Schiele had scandalized the society's "right-thinking" people with their portraits and the eroticism of the naked bodies that stood out on their canvases or on the walls of the galleries; Gustav Mahler had overturned the canons of the traditional symphony, introducing the sounds of cowbells, the poignant notes of a violin and loud trumpet blasts; Sigmund Freud, thanks to his sofa, had opened up the door to all of the nightmares of the psyche.

At the tables of the famous Viennese cafes, discussions were engaged in about the latest exhibitions, the most recent engravings presented in museums, and the theater, of course. Schrödinger loved both the performances that were staged in the small halls and those presented on the great Viennese stages, and admired everything that went on in that world, reveling in the lives of the actors, actresses, and dancers, indeed, always eager to capture the essence of that palpable sensuality. He also loved Schopenhauer's philosophy, that cocktail of Enlightenment, Romanticism and Kantism, amalgamated by the Eastern Buddhist and Hindu doctrines, which the German philosopher had shaped with his captivating and articulate writing, and in which Schrödinger found his questions, the same answers, the same anxieties and suggestions.

He graduated with a degree in physics in May 1910, and, after his military service, he became an assistant at the Institute of Experimental Physics in Vienna, a place that was, however, very far from his true nature and his most genuine scientific interests. To escape boredom, in the summer of 1913, he decided to participate in a meteorological research in the outskirts of Salzburg. Here, the project director introduced him to his children's nanny: Annemarie Bertel, daughter of a photographer from the Habsburg Court, eleven years younger than Erwin. Hence, he began a discreet courtship, made up of bicycle tours, visits to the various local pastry shops and theater rendezvous.

That atmosphere of innocent youth was interrupted by the attack in Sarajevo: the death of Archduke Franz Ferdinand, heir to the throne of Austria-Hungary, and his wife Sofia caused Europe to sink into the abyss of the Great War. Called to the army, Schrödinger was assigned to the western front and, after seeing all of the rawness and stupidity of the war with his own eyes, in 1916, he was finally assigned to a barrack in Prosecco, a village adjacent to Trieste: here, with a surfeit of spare time, he began to study Einstein's theory of general relativity, managing to write a couple of scientific articles on some of the most delicate aspects of the theory.

At the end of the conflict, he re-established contact with that girl that he had met years earlier in Salzburg and, in March 1920, he married her. But it was immediately clear that that marriage would not work: their interests were too far apart, their characters too divergent, their contrasts too heated. If Erwin was very fond of talking about poetry, philosophy or science, Annemarie showed indifference to all of that intellectuality; if Annemarie loved music, Erwin, contrastingly, detested it, to the point that he never allowed a piano to enter the house. The discussions between the two soon began to degenerate. They spoke several times of divorce, but it never materialized, because they agreed that, despite their marriage not corresponding to the romantic ideal of love, it could offer them a safe refuge, a refuge around which other stories could gravitate, other women, other men. Over the years, they both had various affairs: Erwin had three daughters from three different women, none of them his wife; Annemarie had numerous lovers, including Hermann Weyl, a great friend of her husband's. Their marriage then turned into friendship, in an intertwining of confidences and relationships, all lived in great freedom. All of that aside, Annemarie loved her husband, and, while recognizing that it was sometimes impossible for them to live together, she ultimately said that it was well worth it.

A Wave in Search of an Equation

The story of how Erwin Schrödinger arrived at his famous wave equation remains one of the most fascinating pages in the book of science. Arnold Sommerfeld even claimed that this was the most exceptional discovery of all of the exceptional discoveries of the twentieth century. It is an equation that regulates the quantum world, enormously different from our daily world, a world made up of matter and radiation, but on an atomic scale, where the difference between wave and particle disappears and we enter an apparently paradoxical universe. Particles are lumps of matter that exist at a specific point in space–time; waves, however, are scattered throughout space, a garden where paths fork, Jorge Luis Borges would say. So, how can an object exist that simultaneously has the properties of a particle and those of a wave? Yet this is precisely what happens, not only with light, but also with electrons and all other particles. This is what the first pioneers of quantum physics—Planck, Einstein, Bohr, Sommerfeld, de Broglie, etc.—had to take note of with great amazement. This duality, which later proved to be deeply rooted in the laws of the world, had thus added to other singular quantum phenomena discovered over the years, such as the discreteness of energy, the possibility for a neutron or a proton to cross barriers that were apparently impenetrable from a classical point of view. This was the uncertainty principle, with the insurmountable limit that it places on the degree of precision with which we can know, for example, the position and speed of a particle.

It all started in the early twentieth century, when the German physicist Max Planck had proposed a desperate remedy to solve a paradox related to the radiation emitted by a body brought to a certain temperature: he had hypothesized that the light consisted not of a continuous flow of energy, but of a set of particles, photons, whose energy was expressed in terms of a constant—which, today, bears the name of Planck—and their frequency. For a long time, the German physicist considered this discovery to be only a mathematical trick to make things work, a cunning ruse that did not, however, alter his world view. It was thus up to Albert Einstein to show the true face of Planck's constant and the disruptive effects linked to its existence, including the simultaneous corpuscular and undulatory nature of light and the true existence, in nature, of energy quanta. The idea was then taken up and enhanced in an even more surprising way by Niels Bohr, who, starting from the experimental discoveries of Ernest Rutherford, elaborated a model of the hydrogen atom based on the quantization of the electron's orbits around the nucleus. The beauty of this theory was that not only did the electron not radiate during its motion, but, for the first time, the theory was also able to accurately predict the spectral lines associated with the emission and absorption of light by an atom of hydrogen.

With Bohr's proposal, made in 1913, extremely suggestive scenarios opened up in regard to the atomic world and very important questions surfaced about the mysterious nature of quantum phenomena. The light of a given frequency, according to Bohr, was emitted in the presence of quantum leaps between one orbit and another: could it be calculated then, for example, as to when one of these quantum leaps would have occurred? And, ultimately, what did these quantum leaps consist of? How could the idea—somewhat disconcerting—be conceived that the electron suddenly disappeared from one orbit only to appear in another, almost as in a magician's trick? During the 1920s, confusion reigned supreme, physicists groped in the dark. Schrödinger, like many others, was completely dissatisfied with that state of affairs, and his mind refused to give credit to those discontinuities that Bohr and Sommerfeld tried to emphasize: in his opinion, there was a need to bring order into that mishmash; it was necessary to find an idea that gave a reason for the experimental data without the scientists having to sell their souls to the devil.

When enlightenment appeared, it came to a French aristocrat, Duke Louis de Broglie, who had had the opportunity to reflect for a long time on the nature of electromagnetic waves during the First World War, when, in a cabin set up on the top of the Eiffel Tower, he was working as radio operator. It was he who understood that, if light showed wave properties and gave rise to interference phenomena, the same had to happen with all material particles! In other words, de Broglie suggested that there was a wave associated with each lump of matter, whose wavelength decreased as the mass of the particle grew. Following his intuition, the wavelength of an electron was therefore of the same order of magnitude as the distance that separates the atoms in a crystal, so it was easy to obtain experimental confirmation of that crazy idea: it would be sufficient to send a beam of electrons through a crystal and see whether this electronic bombardment gave rise to interference phenomena (such as the sequence of crests and valleys that are observed by throwing two stones into a pond). De Broglie's hypothesis, reported in his doctoral thesis discussed in 1923, found resounding confirmation a few years later in the experiment of two American physicists, Clinton Davisson and Lester Germer. It was a new element from which to start discovering the laws of the atomic world.

In 1925, Erwin Schrödinger was a professor of physics at the University of Zurich when his colleague Peter Debye, a Dutch physicist, deeply perplexed about the nature of de Broglie's waves, proposed that he read the article by the French scientist and present its contents to other faculty members. The seminar was held on December 7, 1925, but on that occasion, Debye's doubts increased rather than diminished: he began to bombard Schrödinger with questions, which inevitably revolved around the nature of these waves, their origin and, above all, the methods of propagation. He publicly launched a challenge to him: "Find the differential equation satisfied by this damned wave!"

Two Different Visions of the World

The equation was found by Schrödinger in the snows of Arosa during the Christmas holidays of 1925, spent in the company of his mysterious lover, an exceptional woman judging by the extraordinary influence that she exerted on his life. It would be sufficient to say that, throughout 1926, Schrödinger produced his best work, publishing about seven times more material than his standard. It was therefore upon his return from his romantic alpine escape, in the first conference at the beginning of the year, that Schrödinger announced his discovery. He began by saying, "Remember that my dear colleague Debye asked for an equation for the waves of de Broglie? Well, I found one!" It was an equation with partial derivatives, obtained thanks to a real stroke of genius, an equation capable of predicting the temporal evolution of a function $\Psi(x)$ defined in all space, i.e., a perturbation that, moving from one point to another, gave rise to the typical phenomena of wave propagation.

Mathematically speaking, this equation was part of a long and fruitful tradition of mathematical physics, the great eighteenth and nineteenth century tradition that had led to the understanding of all of the phenomena related to the vibrations of violin strings and the membranes of the drums, thanks to the Fourier transform and other linear methods. Among its qualities, it was the very nature of Schrödinger's equation that highlighted the existence of quantized solutions! In fact, this equation selected only a discrete number of functions $\Psi(x)$ capable of fully meeting the physical requirements. In so doing, it recalled in full what happens with a violin string, which, when vibrating, presents an integer number of wavelengths, implying the existence of a certain number of zeros, those points of the string that remain stationary over time, while everything else is swinging up and down. Physically acceptable configurations of the violin string are those in which we have, for example, the first harmonic (which corresponds to a wave motion without zeros), the second harmonic (which has only one zero), the third harmonic (which has two), and so on. In music, these are the pure notes emitted by a violin. And, using this analogy literally, Schrödinger showed that he could very elegantly explain the quantized orbits found by Bohr! In fact, they corresponded to those particular trajectories of the electron composed of an integer number of wavelengths of its wave function.

"Wave mechanics": this was the name given to the formulation of quantum mechanics proposed by Schrödinger. Obviously, an explanation was still needed for the nature of the wave function $\Psi(x)$, to understand whatever it represented. Nevertheless, the new formulation fully satisfied Schrödinger's conviction that there are no leaps in nature, that everything is continuous: according to this new point of view, the quantum universe presented itself as an arena full of wave disturbances, sometimes concentrated in restricted regions of space in a manner that gave rise to the illusion of being in the presence of particles. These waves, during their motion, in turn gave rise to constructive or destructive interference with peaks and valleys. It was a seductive vision, in which the phenomenon of the absorption or emission of light by an atom could be interpreted as the continuous change of a violin string plucked by a musician. In addition, it provided the basis for understanding the chemical bonds, the size of the atoms, the relationship between classical and quantum physics, and thus had the additional advantage of greatly expanding the catalogue of possible calculations. Schrödinger's equation therefore immediately aroused significant curiosity and was given great attention: among the most enthusiastic was Einstein, who wrote to him: "The idea of your work is a real stroke of genius"; but Planck also commented: "I read your works as an impatient child, happy to hear of the solution of an enigma that had irritated me for a long time." That equation led to its author's imperishable fame, the most prestigious chair in theoretical physics in Europe—that in Berlin, which had previously been Planck's—and finally, the Nobel Prize in Physics in 1933, shared with Paul Dirac.

The only dissenting voice in the chorus was that of Werner Heisenberg; the two men were not only of different generations, with fifteen years dividing them, but also completely different worldviews. Heisenberg opposed a purely epistemological, philosophical approach to Schrödinger's interest in Greek philosophers and the themes of Hinduism. If Schrödinger's training had taken place with classical texts of mechanics, electromagnetism and thermodynamics, Heisenberg, paradoxically, had first learned atomic physics thanks to the famous physicist Arnold Sommerfeld, and only after classical Newtonian physics. Schrödinger was always looking to capture what he called "the feeling of femininity"; Heisenberg, contrastingly, had a normal married life. Schrödinger left Berlin in 1933, disgusted by the Nazi violence, while Heisenberg was a passionate nationalist, and his involvement in the German atomic bomb project is still a matter of debate. In short, Schrödinger differed in every possible way from his younger colleague, even in physical appearance: Heisenberg had the look of a simple country boy—short hair, bright eyes and an almost tender expression; Schrödinger instead was imposing, with his professorial figure, his immense aura of culture, his broad forehead, his intense gaze behind the rimmed glasses and his original way of dressing.

To his friend Wolfgang Pauli, Heisenberg wrote, in 1926: "The more I reflect on Schrödinger's theory, the more disgusting I find it. What he tries to visualize is probably wrong, pure nonsense." Schrödinger's judgment on his rival's theory was equally trenchant: "A set of repulsive, counterintuitive and abstract formulae." Bohr was on Heisenberg's side. Einstein, however, was on Schrödinger's and, in private, did not hesitate to offer sarcastic judgments: "Heisenberg has laid a large quantum egg. In Göttingen, they believe it; I don't." The scientific dispute between the two focused on the very nature of the new quantum theory. If, for Schrödinger, the atom behaved like a musical instrument, with the different energy levels associated with the various harmonics, for Heisenberg, a coherent view of the world was obtained only by using the algebra of matrices, tables of numbers with precise laws of multiplication and addition. Energy levels were obtained, in this case, by solving algebraic equations, called eigenvalue equations. The difference between the two theories appeared, at first glance, extraordinary: their starting point was different, their presentation was different, the entire mathematical apparatus was different. But, ironically, it was Schrödinger himself who demonstrated their complete equivalence!

It was subsequently Max Born who clarified the mysterious nature of the wave function $\Psi(x)$: it was associated with a probability! So, although Schrödinger had tried in every way to get the random aspects of quantum mechanics out of the door, they had returned through the window. During a visit to Copenhagen, Schrödinger turned to Bohr in a moment of exasperation, saying, "If we have to continue to endure these damned quantum leaps, I'm sorry I ever agreed to deal with quantum theory!" To which Bohr, with a big smile, replied, "We are all grateful that you have taken care of it!"

Further Readings

J. Gribbin, *Erwin Schrödinger and the Quantum Revolution* (Bantam Press, 2014)
W. Moore, *A Life of Erwin Schrödinger* (Cambridge University Press, Cambridge, 1994)

Rasetti. From Atomic Nuclei to Cambrian Trilobites

1938 was the year of racial laws in Italy, the last shameful act of a dictatorship that shortly thereafter dragged Italy, and the whole of humanity, into the immense tragedy of the Second World War. One afternoon, towards the end of that year, upon entering one of the rooms of the Physics Department in via Panisperna, Franco Rasetti found Enrico Fermi and Mario Salvadori engaged in subdued chatter.

"What are you doing? Are you plotting?"

"Salvadori has decided to go to America," replied Fermi gravely.

"Oh really? He leaves, then you do too, and who am I, the idiot who stays behind? I'm going too!"

Shortly thereafter, the famous group in via Panisperna evaporated without leaving a trace behind.

The Pisa years

Franco Rasetti was born in Pozzuolo Umbro, a section of Castiglione sul Lago, on August 10, 1901. His father had an "itinerant" agriculture school at the University of Pisa: in practice, he traveled far and wide through the countryside of Tuscany to teach local farmers how to make the most of the land. He was also an expert in botany, entomology and geology. Rasetti's mother was a lively woman and, although uneducated, was passionate about plants and insects; as Natalia Ginzburg, who knew them well, wrote in the family lexicon, "in that family, they were all entomologists and botanists." Even Franco, from an early age, showed a fascination with insects, about which he knew everything: color, scientific name, order, family. He had a prodigious memory, and when he grew up, he boasted that he could still remember more than fifteen thousand names of fossils and plants. With his father, he set up a singular collection of beetles and specimens of alpine flora. His uncle Gino Galeotti took him climbing in the mountains, and from those expeditions, they came back

© Springer Nature Switzerland AG 2020
G. Mussardo, *The ABC's of Science*,
https://doi.org/10.1007/978-3-030-55169-8_18

down with their backpacks always full of precious material to clean up and classify, such as moss and lichen clods, dead beetles and crystals.

Uncle Gino was another singular figure: a medical pathologist, he took advantage of the presence of his nephew to conduct his curious experiments on the physiological effects of mountain life, exposing him bare-chested on the hairpin bends full of snow so as to measure the drop in his body temperature. To make him appreciate the experience better, he advised his young nephew to read a book on thermodynamics. Years later, Rasetti remembered: "When I finished high school, one of the few concepts that was clear to me was that of entropy. I may not have understood Gauss's theorem well, but I knew thermodynamics to perfection." When it was time to go to university, he enrolled in Engineering at the University of Pisa, the city that the family had moved to a number of years earlier. But after a few months, something happened that completely changed everything: he met Enrico Fermi, at the time a student at the Scuola Normale Superiore. To gain entrance to the Normal, one had to take a very selective entrance exam, and Fermi had amazed all of the members of the commission with his written test, which focused on the distinctive characteristics of sound: he addressed all of the aspects of the problem, displaying a stunning knowledge of the partial derivative equation for waves and implementing methods based on the Fourier series to solve it. It was 1918, and, in the following years, Fermi's fame only increased.

Fermi and Rasetti met while following a course common to the two faculties. Fermi was very shy and reserved and, at first glance, he went unnoticed, while Rasetti had a thunderous and noisy laugh, which immediately attracted attention. If the former was small and stocky, the latter was tall and lanky. However, they had a great intelligence and a keen curiosity towards all scientific questions in common. They later discovered that they both loved the mountains and tennis. A great scientific and human partnership was immediately established between them, laced with humor and seasoned with a great deal of youthful joy: the four years spent together in Pisa were the happiest and most carefree of their lives.

Leaving Palazzo dei Cavalieri, the sixteenth-century building where the Scuola Normale was located, Fermi felt possessed of a great light-heartedness, articulated by his wife Laura Capon Fermi in her delicious book *Atomi in Famiglia*:

> Fermi studied little. He knew much of what was being taught to him and easily remembered what he was hearing for the first time. He had time to devote himself to the fun of goliardic life: he fought with buckets of water on the roofs of Pisa to protect the honor of two bridesmaids that had never been threatened; he engaged in fake duels for reasons unknown to both the challengers despite the consequent frantic scrambling of the cavalry code; he initiated a successful intensive campaign to elect a girl with large feet and male features as the Queen of May, to her considerable embarrassment. It is my opinion that Fermi would not have given body and soul to this kind of life if he had not been dragged there by a new friend, Franco Rasetti.

In the scientific field, Fermi excelled in all theoretical aspects, while Rasetti demonstrated greater skill in the laboratory. If Rasetti read the classics of English literature, Fermi devoured all of the articles on physics published in German in those years: in this way, he learned Einstein's theory of relativity, which appeared in 1916

in "Annalen der Physik", as well as quantum mechanics, and he was not even daunted in the face of the mammoth *Atombau und Spektrallinien* by Arnold Sommerfeld. He shared all of this with his friend, who absorbed it like a sponge (although Rasetti later commented, "The only thing I regret is that I managed to absorb only the smallest part of his knowledge"). In 1920, Rasetti finally decided to leave his engineering studies to enroll in a course in physics: his challenge was to demonstrate that he would also be successful in this field.

Spectroscopy of a Friendship

Franco Rasetti was a great individualist, refractory towards rules and alien to discipline. He reacted to formal situations in a sardonic and detached way. He established relationships exclusively with those whom he believed to be his peers. A vivid portrait of his character emerges once again from Laura Fermi's words: "He was a silent young man, the girls were attracted to him, even though, or perhaps because, he did not seem to notice. He looked at them with cold objectivity, meanwhile forming his mouth into a smile that was both amused and ironic at the same time." From the time he was a boy, he had an incredible versatility and a strong interest in thousands of things. It wasn't just the laboratory bench: he said that life was too big for a single cat, by which he meant physics. There were the mountains, with their breathtaking views, to be enjoyed after a climb made with nails and crampons, braving the wind and the cold of the most remote peaks of the Alps. In those remote corners, he heard the echo of the valleys and the babbling of the streams, and enjoyed the pleasure of the sparkling air. There was botany, with its herbarium and the infinite leaves that he collected and classified in a painstaking way in his notebooks, which also contained a thousand color drawings. There were his descents into caves, and the corresponding collection of minerals. There was chemistry, geology. There was the search for exciting pastimes through which to give vent to his restless intelligence, the establishment of an "Anti-neighbor Society," dedicated to putting a little salt in the life of his fellow men, for example, by designing jackets that were buttoned with small iron padlocks, so that you could not take them off without the assistance of a blacksmith.

Not everyone appreciated his exuberant personality: "At the table, he spoke incessantly, but always about physics, or geology or beetles. He is very intelligent, my father said of him, but he is arid! It is very arid! He never talks about politics, he doesn't care," Natalia Ginzburg wrote of him. Paola Levi, Ginzburg's sister, added that he was a "perfect pure mineral," in other words, he had extraordinary intellectual interests, but little interest in humans. All true, but with Fermi, it was another story. The two of them huddled together on the Apuan Alps, Rasetti's long, pale legs climbing over holes and crevasses before breaking into a very fast run so as to chase a butterfly, with Fermi have no trouble keeping up with him. Upon returning, Fermi would spend time in the Rasetti household, where he was welcomed with familial warmth, almost like a son. His friend's mother was very fond of the boy, so laid back, with a calm air.

The years passed quickly, and at the beginning of 1922, Rasetti found himself facing his thesis work, which focused on alkaline gas vapors and the light that they emitted. The research fell within the sphere of spectroscopy, a science that had already taken hold in the nineteenth century, but that showed its greatest potential in the early decades of the twentieth century, when the first revolutionary ideas proposed by Bohr, Sommerfeld and Planck had sprung up as to how to interpret those sharp dark lines obtained by placing an ampoule full of gas next to a light source. For Rasetti, spectroscopy would turn out to be a longtime travelling companion during his scientific career. He had started his thesis work on optical benches on the ground floor of the historic Palazzo Matteucci, in Piazza Torcelli, where the Institute of Physics was housed. Together with Enrico Fermi and Nello Carrara, however, they had long been granted permission by Professor Puccianti to access the other laboratories of the University of Pisa, and the three had improvised new experiments there, rescuing the instrumentation from the layers of dust in which it was covered and, after obtaining induction coils and generators of very high electric currents, even building the tubes themselves to produce X-rays. Rasetti graduated in November 1922, while Fermi had already graduated in June, after presenting a thesis on Röntgen rays. Fermi had illustrated his work with monotonous competence in front of the dull gaze of the members of the commission, receiving praise in the end, but Rasetti doubted that any of them had truly understood what had been discussed. 1922 was not only the year of their graduation theses: in October, Benito Mussolini, leading about 25,000 black shirts of the Fascist National Party, marched on Rome, and the king, out of fear that the situation would degenerate, turned governmental power over to them. Fermi understood immediately that this change of power would be disastrous for democracy and science in Italy.

Rasetti, on the other hand, prey to his impulsiveness as usual, had participated in the march on Rome, believing it to be just another student group like any other. His adhesion to fascism was truly an improvisation, a half-clowning: as Natalia Ginzburg's father Giuseppe Levi had said, Rasetti had no interest in politics, and the Matteotti murder, in June 1924, alienated him from any sympathy he could ever have felt for the fascist regime. After graduation, the paths of the two friends divided, at least momentarily: Rasetti was called by the professor of Physics Antonio Garbasso to fill the role of assistant at the University of Florence, while Fermi instead went first to Göttingen, and then to Leiden[?]. For Fermi, it was perhaps the worst time of his academic life: in Gottingen, he was treated with an air of superiority by that somewhat snobbish group. Fermi was too pragmatic in spirit for those such as Heisenberg, Born, Pascual or Pauli, lost behind the rules of commutation and the abstract formulations of physical laws; he was a "quantum engineer," rather than a "quantum mathematician." It went a little better for him in Leiden; in fact, he spent some pleasant months in Paul Ehrenfest's group. In 1924, however, again due to Garbasso's interest, Fermi was also called to Florence, to a chair of mathematical physics, and so, fortunately, the partnership between the two was recomposed, laying the foundations for what would become the famous group of via Panisperna.

Those were extremely profitable years that left a lasting trace on the development of physics. Rasetti concentrated on issues of spectroscopy, in those years, the most powerful investigative medium on the physical behavior of matter; Fermi instead meditated more deeply on quantum mechanics. When Schrödinger's famous article was published at the beginning of 1926 on the wave formulation of quantum mechanics, it was truly liberating for Fermi: he threw away all of that complicated formalism that came from Heisenberg and his collaborators, and went headlong in that new formulation to produce some results on the polarization of light or on the dynamics of a set of particles. And it was precisely through reasoning on the properties of the wave functions relating to many particles and to Pauli's exclusion principle that, in March 1926, Fermi came to a new and fundamental statistical law applied to particles, which, from then on, were called "fermions". His statistics later proved to be fundamental for understanding the behavior of electrons in metals, for designing transistors and for understanding the laws governing neutron stars. With this discovery, Fermi acquired a remarkable international reputation.

Fermi and Rasetti formed a formidable team, and towards the end of the 1920s, the only embarrassment they suffered was an embarrassment of choice as to which themes they would work on. They first studied the electron spin, that degree of internal freedom identified by the experiment by Uhlenbeck and Goudsmit; then, they went on to study the effects of a magnetic field on the resonance radiation, realizing the equipment that they needed themselves, with diodes and ammeters recovered in the workshop.

A Famous Photo

Orso Mario Corbino had always had a good nose: he was an experimental physicist of Sicilian origin, he graduated from Palermo, a student of Augusto Righi, and, since 1908, he had taught at the University of Rome. For many years, he had cleverly entered the vortex of politics and had it in mind to revive Italian science: it was child's play to establish the first chair of theoretical physics in Italy and assign it to Enrico Fermi. To create an international school of physics, he also called Franco Rasetti, knowing that Fermi entrusted the direction of each delicate and demanding experiment to "his arm," his most expert collaborator. Corbino went to great lengths to publicize the birth of this group, and there was soon a host of students eager to embark on that adventure: Emilio Segrè, Ettore Majorana and Edoardo Amaldi. Bruno Pontecorvo also came from Pisa later.

The Institute of Physics in Rome was, at the time, in via Panisperna, at number 89. The classrooms in that austere building, hitherto empty and lonely, suddenly began to come to life: there were a thousand voices in the corridors, Rasetti's shrill laughter, the coming and going of new experimental equipment, which then found a place in the old basements. In the group, a cheerful goliardic atmosphere was immediately established, resulting in a string of ecclesiastical nicknames: if Fermi was the "Pope", because of his infallibility, Corbino, his direct superior, was the "Eternal Father";

Rasetti, as Fermi's right-hand man, was obviously the "Cardinal Vicar" or the "Venerated Master"; Segrè, for his bad temper, was the "Basilisk", the legendary reptile capable of killing with its eyes; Majorana, who criticized everything and everyone, was the "Grand Inquisitor"; Amaldi was the "Abbot", while Pontecorvo, the latest arrival, was the "Cucciolo"; Enrico Persico, who taught in Florence, and was considered in partibus infidelium, was the "Cardinal of Propaganda Fide"; finally, Giulio Cesare Trabacchi, chemist at the Istituto Superiore di Sanità, who, with his knowledge of the shops and pharmacies of Trastevere, supplied the materials for the experiments to the young physicists of the group, was called the "Divine Providence."

Segrè said that everyone in the group began to speak with the same Pisan cadence: Fermi had a deep voice and engaged in a particular form of interplay, imitated a bit by everyone; Rasetti infected the others with the stories of his English readings, his travels, the culinary rites that he had seen around the world. Fermi's interests embraced all physics, and he mapped them out freely on the blackboard as if he were in class, clearly, consequentially, without any logical leaps. He had the forward drive of a steamroller. He preferred concrete problems and was wary of excessive amounts of abstraction in theories, which, in any case, he reworked in his own way. So it was, for example, with the theory of quantum fields, about which he wrote a review article that constituted the reference point on the subject for generations to come. He devoted a lot of time to his students, teaching them all of the techniques for dealing with a question or the mathematical tricks for simplifying it. Rasetti was also generous in sharing his experience and laboratory tools with the young researchers, although some found it difficult to work with him because of his irregular hours and small manias. One of the latter was that the instruments had to satisfy a criterion of "beauty." One afternoon—Pontecorvo recounts—he had a curious exchange of jokes with Fermi, accusing him of having made an ugly instrument: "But it works," said Fermi. Then, Rasetti, unnerved, broke out hysterically: "In experimental work, you are capable of performing truly unworthy acts. Look at Edelman's electrometer [Rasetti meant that gleaming, splendid chromed and elegant device, which, in his eyes, was the symbol of technical perfection]; if you were convinced that, to get some data, you had to sprinkle it with "chicken blood", you would do it immediately. I, on the other hand, would not be able to perform such an act, even if I knew it could allow me to win the Nobel Prize."

To improve the group's competence in the experimental field, in 1929, Segrè went to Zeeman in Amsterdam, Amaldi visited Debye in Leipzig, and Rasetti went to Pasadena's Caltech to work with Millikan and, on his return, did nothing but praise the beauty of California: the orange groves of Pasadena, the sweetness of the Los Angeles hills, with all of those eucalyptus trees. At Caltech, thanks to some measurements on nitrogen molecules, he had conducted a very clever experiment, with which he had put the model that proclaimed the nucleus to consist of protons and electrons in crisis (however, we had to wait until 1932, for the discovery of the neutron made by Chadwick, to completely solve the mystery). Later, when research interests had shifted in the field of nuclear physics, Rasetti went to learn the fundamentals directly from Lise Meitner, in Berlin. The group in via Panisperna also benefited from fruitful exchanges with many foreign researchers, who came to Rome to speak with Fermi and to find out about the latest news: scientists of the caliber of Hans Bethe, Felix Bloch, Edward Teller, and many others.

When the first news of the experiments conducted in Paris by the Joliot-Curies concerning the production of new radioactive isotopes obtained by bombarding the elements with α particles arrived in Rome, Fermi immediately understood that the neutron would be much more useful as a test particle, because neutrons could penetrate the atomic nucleus without undergoing Coulombian repulsion by protons. He immediately suggested to Rasetti that they use neutrons emitted by a polonium source, but the experiment did not produce the desired results, because, it was understood later, the energy of the neutrons used was not sufficiently large. At that point, Rasetti, following his well-known philosophy that physics cannot exhaust the richness of life, left for a long trip to Morocco, where he intended to explore the mountainous areas and study the country's flora and fauna. Fermi, in Rome, initially continued his experiments alone, and then involved Segrè and Amaldi. He intended to use neutrons to systematically bomb all of the elements present in the periodic table, starting from those with the lowest atomic number and going up to uranium. It soon became clear that a chemist's help was needed. Oscar D'Agostino joined the group, returning fresh from the Joliot-Curie group in Paris. Having obtained the first encouraging results with the bombing of fluorine, Fermi telegraphed to his friend in Morocco, begging him to return as soon as possible, which, in his time, Rasetti actually did. On the terrace of the Institute of Physics, one afternoon in 1934, one of the most iconic images of science was taken by Bruno Pontecorvo, portraying five of the "boys of via Panisperna": Fermi, Rasetti, Amaldi, Segrè and D'Agostino. Fermi is the only one wearing a tie.

In a few months, 60 elements were irradiated and 44 new nuclides were discovered. In 1934, they came to the discovery—with a good dose of luck and thanks to Fermi's exceptional intuition—of a sensational phenomenon, which contributed to changing the fate of the world. The bombardment with neutrons seemed to work, but the results were fluctuating: they changed according to whether the experiments were carried out on a wooden or a marble table. Rasetti at first got angry with Amaldi and Pontecorvo, especially with the latter, thinking that those embarrassing series of measures were due to his inexperience, to the point that he sent him out of the laboratory! Poor Pontecorvo wandered, disheartened, for several days in the corridors, after which he was allowed back into the laboratory, because the same funny phenomenon repeated, even in his absence. The first attempts to understand its origin were unsuccessful, until, on the morning of October 22, 1934, when, with an intuition that became legendary, Fermi had the idea of interposing a paraffin screen between the neutrons and the silver cylinder to radiate. The Geiger counter began to crackle like a machine gun, recording an increase in artificial radioactivity that had never been achieved before. Since paraffin contains hydrogen atoms, they were the ones to slow down the neutrons before they reached the nucleus of the elements to be irradiated, increasing their penetration capacity. Other frantic checks were immediately made: the source and the target were immersed in the goldfish fountain that was in the middle of the Institute's garden and, in this case as well, the Geiger counters went crazy. On the evening of that same day, everyone gathered at Amaldi's home to write an article to be sent to the magazine "La Ricerca Scientifica". Strongly urged by Corbino, Fermi and the others of the group made a patent application shortly afterwards for the method of production of radioisotopes through the use of slow neutrons.

During the subsequent studies, the boys in via Panisperna observed that, in the case of the heaviest element known at the time, uranium, the bombardment of neutrons gave rise to a radioactive body that they could not chemically identify. Mistakenly, Fermi thought that he had created transuranic elements, but, in reality, they had witnessed, for the first time, a phenomenon of nuclear fission, in a nutshell, the process behind the atomic bomb. After a few years, it was Lise Meitner and Otto Frisch's turn to clear up this mystery and open up the doors to a new enigmatic world.

The Great Refusal

In December 1938, taking advantage of the delivery of the Nobel prize that he had won precisely for the discovery of the phenomenon of slow neutrons, Fermi left Italy with his wife and two children, heading for Stockholm. Shortly thereafter, he landed in the United States, and here, behind the western stands of the abandoned Alonzo Stagg Field stadium of the University of Chicago, on December 2, 1942, he carried out the first self-powered artificial chain reaction of an atomic battery. The Manhattan Project started the following year.

Even Rasetti emigrated abroad, bidding farewell to Italy on July 2, 1939, when he embarked from Naples on the ship Vulcania to New York. They had offered him a chair at the University of Laval, Canada, where they had given him the task of creating a Department of Physics from scratch. Despite the scarcity of means and funds, Rasetti continued to work on his own on very refined experiments involving cosmic rays: building himself all of the necessary tools—mass spectrographs, ionization chambers, Geiger counters and devices to implement the method of coincidences— he measured the average lifespan of muons, wrongly identified at the time with the nuclear particles hypothesized by Yukawa. This measure, also done independently by Norris Nereson and Bruno Rossi—another Italian who fled to America due to racial laws—provided the first direct verification of the relativistic law of time dilation: in fact, traveling at speeds close to that of light, the average lifespan of these particles was considerably dilated, compared to that in their resting system, meaning that they could safely reach the ground after being produced in the upper atmosphere.

Like many other nuclear physicists of the time, in January 1943, Rasetti was contacted to join the Manhattan Project, that gigantic program set up by the United States to build the first atomic bomb. Rasetti refused. His motivation was simple and disarming: the stupidity of war. He later wrote in his diary:

> There are few decisions that I have ever made in my life for which I have had less regret. I was convinced that nothing good could come from new and more monstrous weapons of destruction, and subsequent events fully confirmed my suspicions. As perverse as the Axis powers were, it was clear that the other front was sinking to a similar moral (or immoral)

level in the conduct of the war, as evidenced by the massacre of 200,000 Japanese civilians in Hiroshima and Nagasaki.

This stance later prompted him to sharply condemn those who had taken part in it, his friend Fermi for starters, and ultimately to abandon physics so as to devote his time to a study of Cambrian trilobites and the classification of alpine flora. For his new research, the National Academy of Sciences awarded him the prestigious Charles Doolittle Walcott Medal in 1953. He became an authority on the subject, a celebrated discoverer of rich fossil deposits and author of fundamental books on botany and paleontology, such as the fascinating volume *I fiori delle Alpi*, published in 1980 by the Accademia dei Lincei.

The reasons for his choice were expressed in a letter to his friend Enrico Persico, written in the immediate post-war period:

> I was so disgusted by the latest applications of physics (with which, if God wills, I managed to have nothing to do) that I seriously had to think about dealing only with geology and biology. Not only do I find the use that has been made and is being made of the applications of physics monstrous, but, moreover, the current situation makes it impossible to return to this science that once had a free and international character but that now is only a means of political and military oppression. It seems almost impossible that people whom I once considered to be endowed with a sense of human dignity could lend themselves to being the instrument of these monstrous degenerations. Yet, it is really so, and it seems that they do not notice it. Of all the disgusting shows these days, there are few that match that of physicists working in laboratories under military surveillance to prepare more violent means of destruction for the next war.

Words like stones, which have made Rasetti one of the most emblematic figures in the history of science, a scientist capable of setting a limit upon his research activity in the name of social and civil responsibility. Words, moreover, that are still very timely, given the "stupid choice," as Rasetti would say, of many governments who try to solve the world's problems by dropping "smart" bombs and handsomely financing those groups of scientists to build them. We therefore like to remember Rasetti as that eclectic man who collected insects in cigarette boxes, who passionately loved life and nature. "I must admit," he wrote, "that discovering Nature's secrets is among the most fascinating things there can be. But it may be that something is both very fascinating and very dangerous … I think that men should ask themselves more deeply about the ethical motivations of their actions. And scientists, I'm sorry to say, don't do it very often."

Further Reading

L. Fermi, *Atoms in the Family. My Life with Enrico Fermi* (The University of Chicago Press, 1954)

Spallanzani. Priest Jokes

In the summer of 1793, the small town of Scandiano, at the foot of the Apennines of Reggio Emilia, was the scene of a singular event that soon created turmoil not only in that small Emilian village, but also in the entire scientific world of the time, leading to a frenzy of heated controversies and animated discussions. The protagonist of this affair was a lanky sixty-year-old abbot, with a spacious forehead and lively black eyes, who, during that summer, discovered a strange passion for bats. In the medieval fortress of the Boiardo, on top of the hill overlooking the town, there were entire bat colonies, and he spent his days spying on them while they hung upside down in clusters in basements and attics. However, after a while, he began to feel bored with this mere "birdwatching", and it occurred to him that he could equip himself with a large net and begin actively hunting for the flying mouse, assisted in this crazy enterprise by his brother and a nephew. And, indeed, several unfortunate bats did become entangled in the nets of the trio led by that diabolic abbot; the unfortunate little beasts, placed in a number of baskets, were brought inside the castle and ultimately released into a room. The rediscovered sense of freedom of the poor animals manifested itself in the very rapid fluttering of their membranous cloaks, as they incessantly flew back and forth between one wall and another, taking sudden twists and turns throughout the vaults of the hall, their high-pitched screeches echoing everywhere. Sadly, it was a rather ephemeral freedom, given the cruel fate that awaited them, as they were soon to be subjects of a particular experiment…

A Priest Sui Generis

"Experimenting" is the right word to use in regard to the first authentic modern biologist and physiologist, Abbot Lazzaro Spallanzani. Truly a priest sui generis, free from both philosophical bonds and religious beliefs, attracted as much by the

© Springer Nature Switzerland AG 2020   179
G. Mussardo, *The ABC's of Science*,
https://doi.org/10.1007/978-3-030-55169-8_19

mystery of life as the myriad forms it takes. The ecclesiastical path, for a wealthy young man like himself, had perhaps been a forced choice, a way like any other to free himself from the boring problems of everyday life. Born in Scandiano in 1729, he began wearing the cassock at the age of eight, and at twelve, he had his first tonsure, definitively taking the vows in 1763. As it soon became understood, he cared little about his clerical office; he officiated in an irregular manner and never participated in any other rites, such as baptisms, marriages or extreme anointings: for him, the assistance and moral protection of the Church were only a means to continue his research undisturbed and to satisfy his thirst for knowledge. After completing his grammar and jurisprudence studies in 1754, according to his father's wishes, from that year on—thanks mostly to the influence of his cousin Laura Bassi, the first woman to graduate from Bologna—he devoted himself exclusively to science, studying astronomy, geology, mathematics and botany, and quickly gaining a great reputation throughout Europe for his highly original naturalistic research. This fame was sanctioned by his election as a member of the Royal Society of London in 1768, and by his appointment to the chair of Natural History at the University of Pavia in 1769, a prestigious academic position that he held for the following thirty years.

A tireless traveler and a very keen naturalist, Spallanzani was, above all, a passionate experimenter, always ready to conduct the most daring experiments for the pure joy of knowledge and for the mere cult of truth. There is no chapter in the natural sciences, in fact, that does not bear a trace of his work—from the phenomena of hibernation to those of breathing, from the processes of digestion to artificial reproduction. In each of these investigations, he had the intelligence and resourcefulness of an unparalleled pioneer. In the summer of 1761, he undertook the first of many scientific journeys that would take him to various parts of Italy and abroad, as a natural historian, a collector of finds and, in general, an observer of the world's many offerings: he headed towards the Apennines of Reggio Emilia, towards the lake of Ventasso, curious to discover the sources of the rivers that came down from the mountains and, ultimately, their origin. In fact, a debate had been going on for some time, setting the famous French philosopher Descartes in opposition to the Italian naturalist Vallisneri. The former claimed that the water of the rivers had a marine origin, that the lack of salt was due to its purification as it percolated upwards through the various layers of soil. The thesis of the latter, however, was that the water that gushed from the mountain springs originated from the rains and snow; instead of reaching the bowels of the Earth, it accumulated in various cavities thanks to the impermeability of some rocks. It was necessary to determine how to resolve the controversy once and for all. To verify that Vallisneri's hypothesis was actually the correct one, as his instinct suggested, Spallanzani related the number and the flow of the springs to the average annual rainfall in the area and measured the presence of the various salts in the waters, comparing them with those found in rocks and finding a strong correlation. Furthermore, he could not resist the temptation to dispel a local myth, according to which there was a great vortex at the bottom of the lake: armed with a very long pole, he probed the depths of the waters at every point, finding nothing but a seabed without cavities.

A Head Problem and a Stomach Problem

His experiments on the spontaneous regeneration of the amputated tails, limbs and heads in earthworms and the tadpoles of amphibians, snails and salamanders—conducted during the three years prior to his election to the Royal Society of 1768—had toured the courts of Europe, arousing great curiosity, great discussions among the wise men of the time and a real massacre of the poor gastropods. In fact, there were numerous naturalists, as well as those interested only in the profane, who wanted to repeat the experience. It was easily reproducible and did not require particular dexterity: the fashion of decapitating snails spread like an epidemic, and it was not a good time for the unfortunate mollusks. There was also a famous lawyer who rose to take up their defense, arguing against Spallanzani: "We are nothing less than stunned at these cruel executions ... these more or less skilled executioners are willing to destroy the species to find out about the properties that characterize it."

The subtle implications concerning the fate of the snail's soul, once its head had been detached and before its re-growth, had given rise to extremely violent controversies among philosophers of the caliber of Voltaire and equally prominent religious figures: what happens to it, questioned the sage, when the animal is reduced to pieces by the experimenter? If it is the soul that gives life to the animal and this soul is indivisible—many argued—how is it possible that every other part of the snail will preserve life? Would the soul therefore be divisible, sectionable? Or does it temporarily take refuge in one of the pieces of the animal? There was abundant material for discussion and controversy. And Voltaire was a champion in this! In a letter to the Reverend Father Élie-Catherine Fréron, with whom he had entered into a tense epistolary conflict, he explained his reasoning, which came out more like a slap in the face: "Philosophers and theologians think like children. Who will be able to explain to me how a soul, that is, a principle of sensations and ideas, can reside between four horns, and how does this soul remain in the animal when the four horns have been cut? This surprising object of our confused curiosity is connected to the problem of nature, of the first principles of things [...]. No matter how little you dig, you find an infinite abyss."

Yet, Spallanzani was always there to dig, to try to bring light into that infinite abyss of ignorance. His research on rotifers and tardigrades (strange animals varying in size from a few fractions of a millimeter to a millimeter, which populate the sands on roofs and the moss on walls) was surprising. Left to dry, these beings deform and shrink completely, reducing themselves to an amorphous mass in which they can remain for very long periods without exhibiting any sign of life. However, if rehydrated, they come back to life perfectly, in a sort of "resurrection", an almost blasphemous word in the mouth of a Catholic priest. According to Spallanzani, it was only necessary to study the phenomenon experimentally, thanks to perfectly controllable conditions, which he did, with spectacular incisiveness, spending long hours with his eye fixed on the microscope, ultimately demonstrating, irrefutably, that the phenomenon was essentially an extreme form of animal hibernation.

Spallanzani's fanaticism for experimentation and the desire to know were so extreme as to make him overcome any repugnance, to the point of transforming his own organism into one of purely experimental means. For example, to study the digestive process, about which, at that time, very little was known (there were those who claimed it was due to mechanical contractions of the intestine and those who instead favored the chemical action of gastric juices), he did not hesitate to swallow a metal cap full of breadcrumbs with punctures in its walls, tied to a string. Being incompressible, the capsule was obviously insensitive to the contractions of the intestinal walls, and any digestion of its contents would therefore have been attributable to the action of the gastric juices. Retrieving the capsule by pulling the string up his throat, Spallanzani noted that the inside was filled with a gelatinous substance and that very little remained of the original bullet of bread. It was the first step towards understanding the nature of the digestive process, which did, in fact, seem to originate from the chemical action of gastric juices. To definitively prove this conjecture, he had the idea of obtaining a certain quantity of gastric juices directly from his own vomit. He collected the acid samples in glass bowls, added pieces of cooked beef to them, closed the containers and waited for thirty hours, discovering, upon opening them, that the meat was completely brittle and reduced to a pulp: through this direct experience of "Artificial digestion", he finally solved the riddle.

By deepening the studies on tissue generation, in 1776, Spallanzani came to discover the animal nature of spermatozoa and the influence of temperature on their vitality and mobility. In the following two years, he devoted himself exclusively to the problem of reproduction, the vital process par excellence, obtaining, in 1777, the first "artificial fertilization" in history, using frog and roe eggs, and then successfully repeating the experiment on a dog.

The triumph of his ideas brought him great fame: he was respected by the French Académie des Sciences, honored by Frederick the Great and Joseph II, and by Albrecht von Haller and Galvani, who dedicated their most important works to him. In response, Spallanzani took advantage of this to solicit salary increases. He went so far as to ask the right to celebrate a large number of masses, and to use the money collected during these for his experiments, which were very expensive, since they required a large number of ampoules, test tubes and amounts of various materials.

The Spallanzani Affair

Due to his uncompromising character and his polemic vis, Spallanzani had made several enemies within the University of Pavia, and some of his colleagues hatched a full-fledged conspiracy against him between 1785 and 1786. During those years, Spallanzani had undertaken a long journey through the area of ancient Constantinople: he had deepened understanding of the geology of the Bosporus coasts, collecting crystals from the Armenian plateau and southern Anatolia; he had studied the local marine biology, the various species of birds, and he had also been interested, with a detached scientific spirit, in the customs of the various people who animated the life of the town that had once been one of the world's most glorious capitals. Upon his return, however, a bitter surprise awaited him.

Upon his arrival in Vienna, the pleasant stop that he had thought to enjoy in the company of the Prince of Kaunitz-Rietberg and other members of the Imperial Court soon turned into a nightmare. Indeed, he came to learn from the Count of Cobenzl that he had been officially accused of having stolen several items from the Natural History Museum of the University of Pavia for the purpose of enriching his personal collection, kept in Scandiano. But even more shocking, he learned that the news was widely known in all circles of the city, spread, above all, by a guard who worked for his Imperial Majesty in Vienna, and who also happened to be the brother of one of his accusers, Gregorio Fontana. Fontana, a mathematician from Pavia and also a religious man, was not the only one implicated in this story: in fact, the canon Serafino Volta, custodian of the Pavia museum, the naturalist Giovanni Antonio Scopoli, and Antonio Scarpa, who taught Anatomy and Surgery at the same university, were also involved. Everyone, in one way or another, had a pending account with Spallanzani, and they had vowed to make him pay for it: some for having been treated carelessly, some for their different vision of the natural sciences, some for pure and simple academic envy.

The affaire started with Serafino Volta. Taking advantage of Spallanzani's absence, he had made an inventory of the naturalistic specimens present in the museum, noting, according to him, the disappearance of several items. Since it was known that, thanks to his extensive travels and acquisitions over the years, Spallanzani had set up a private collection in his native town, Volta visited the house at Scandiano in the guise of a foreign scholar, and he spent some time browsing among the collections of shells, stuffed animals, quartz crystals, snakes and scorpions preserved under glass. In his complaint to the rector of the University of Pavia, he wrote: "Having made a visit to the Scandia museum in incognito, it was clear that all its five rooms contain thefts made at the Pavia museum. I realize that the value of these thefts amounts to a few thousand scudi and that, for some of them, there is no price." Had the accusation proven true, it would have been the end of the great naturalist acclaimed by academies around the world, revered as a teacher by his students and the Courts.

Spallanzani immediately proclaimed his innocence to the Habsburg authorities and, returning to Pavia, devoted himself to writing his defensive memory, going on the counterattack and destroying, one by one, all of the accusations that had been made against him. But, as the Barber of Seville sang, "Slander is a breeze, a very gentle aura that is insensitive, subtle, slightly, gently begins to whisper. In a low voice, hissing, it is flowing, it is buzzing … At the end, it overflows and bursts, spreads, doubles and produces an explosion, like a cannon shot." This noise was fueled, in a subtle but extremely incisive way, by the four conspirators: meeting one night in Father Fontana's study at the Ghislieri College of Pavia, thanks to valuable tips from Scopoli, they wrote a letter full of gall and false accusations that was copied dozens of times by one of Fontana's students and sent to the four corners of the world. In fact, Spallanzani learned that the terrible letter had reached the academies of Paris and Gottingen, the Royal Society of London, the Stockholm Academy, even the mathematician Jakob Bernoulli II of the Petersburg Academy, who, when he wrote to ask for an explanation from one of his colleagues in Berlin, received a reply that the latter had received a similar letter. The scandal soon became public knowledge.

Fortunately, the Lombardy government officially opened an investigation, giving a mandate to two renowned scientists to go to the two museums and check whether the complaint made by Volta was founded. After a rigorous examination of the various accusers and the numerous witnesses, with a sentence issued on May 26, 1787, and made official with an imperial decree a few months later, the Court fully acquitted Spallanzani, dismissing Volta in the main body of the text and expressing a harsh reprimend against the other accusers.

For Spallanzani, it was the end of a nightmare. He printed the text of the imperial decree at his own expense and sent it to all of the academies in Europe, to scientific circles, friends, various universities. But he didn't just limit himself to this: he wanted to take ferocious revenge against Scopoli, who claimed to excel in the art of classifying plant and animal species. One day, Scopoli received a strange package: it contained an odd worm and, as he wrote, "this wonderful worm, an animal never seen or described before … was vomited up on February 25, 1784, in Piedmont by the wife of Mr. Vincenzo Domenico Grandi […] six hours before she gave birth to their child." Or so, at least, he was told. From the analysis he made of it, the verdict seemed unequivocal: it was an intestinal worm, but of a very singular type and never seen before. He named it Physis intestinalis, and it became the subject of a detailed table published in his book dedicated to flora and fauna: "I ask the reader to give a good welcome to my worm: and since it does not resemble any other kind of intestinal worm, I have given it the beautiful name of Physis from the Greek word meaning bladder, since the hollow and dilated body of this worm resembles a bladder." Excited by what he believed to be a great discovery, Scopoli even dedicated it to the president of the Royal Society of London, Sir Joseph Banks. In reality, it was a strange artifact formed by joining the trachea, esophagus and part of the goiter of a rooster. Scopoli had fallen into the trap! The grave mistake was immediately noticed, causing serious damage to Scopoli's fame, and also his health, as he died shortly thereafter. In fact, with the signature of Dr. Francesco Lombardini, two pamphlets had appeared in which not only Scopoli and the worm, but also Scarpa and Volta, were ridiculed. Spallanzani always denied it, but everyone knew that he was author of both the joke and the letters that had exposed it.

See with Your Ears

His polemical streak, his sharpness of judgment and his mania for experimentation accompanied Spallanzani with the same indomitable energy until the last years of his life. And there he was, old but as agile as a hare, wallowing in the rooms of the Boiardo fortress to try to understand how the magical bats managed to orient themselves, even when there was very little light. Once the candles were out, he discovered that, while the owls that he used previously had crashed into the walls, falling to the ground like rags, the bats continued to fly undisturbed, dodging all of the obstacles that he had diabolically placed in their way. Thus began an enthralling melee between the flying mice and the sprightly scientist with a long black cloak, so symbiotically similar to his victims.

To explain their mephistophelic orientation skills, he initially conjectured that there was residual light in the room, visible only to bats. Although plausible, the hypothesis was immediately dismissed when he covered their eyes with balls of mistletoe, even going so far as to eventually blind them completely: in fact, even without their eyeballs, those poor beasts were able to twirl around the threads stretched between the walls, hiding themselves, without fail, in the ceiling grooves. This was a true act of devilry, with which, however, it was necessary to come to terms, at any cost. He then blinded some swallows to understand whether these animals could also fly without eyes, but the outcome was negative. What, then, was the mysterious sensory organ used by bats? The sense of smell? The sense of touch? With merciless cruelty, he refined his means of investigation: excluding the sense of smell by blocking their senses with resin, and excluding touch by covering their entire bodies with a very thick paint. Nothing. The bats continued to fly undisturbed. Finally, as he wrote in his Journal of Bats—a detailed account of his research—the turning point came when he poured wax from a candle into the large ears of one of the small mammals. It was only then that he noticed the total bewilderment of the unfortunate beast, which, trying to fly, would crash clumsily into the walls or other obstacles. The bats thus saw with their ears! "The phenomenon," he wrote in the Journal of Bats, "can be explained by saying that the movement of the wings and of the body, hits the air and the latter bounces off the surrounding solid bodies, which the bat feels and avoids." The true explanation of the phenomenon, as is now known, is a bit more complicated than that elaborated by the acute Spallanzani: the bats emit ultrasounds, which, reflected by the objects that they find along the way, are perceived as echoes by the animal's large auricles. The shrieks emitted by the bat are very short, on the order of a hundredth of a second, and since both ears intervene in the location of the obstacle, the distance is assessed thanks to stereoscopic hearing. In order for the emitted sound not to cover the echo, the bat, in the instant in which it emits the shriek, closes its ears through the contraction of a muscle and reopens them immediately afterwards, becoming automatically sensitive only to the reflected sound: a phenomenon of natural adaptation of rare effectiveness and complexity. A joke of nature, one might say, a real priest joke.

Further Reading

R. Dawkins, *The Blind Watchmaker* (W.W. Norton & Company, New York, 1996)

Touschek. The Lord of the Rings

"It is difficult to describe Bruno Touschek if you have not met him. In contrast to the stereotype of the Physics professor, absorbed in his thoughts and a little absent-minded, you have to imagine a restless type, inclined to make eccentric and extravagant jokes, to play hybrid word games (Austrian + Italian + English) and to respond the address of his interlocutor with bright, darting eyes. Bruno was a lover of Karl Kraus and his Central European satire, a jewel of sarcastic literature; moreover, Bruno knew how to draw, with that merciless trait that we can see in certain drawings by Egon Schiele."

This is how Carlo Bernardini described him: Bernardini knew him very well and had lived alongside him for years. They were companions in a very special scientific adventure, which was a real challenge, both theoretically and technologically, and that took place in the 1960s around the enchanting hills outside of Rome, in Frascati, in the newly born laboratories of the National Institute for Nuclear Physics.

In the Tiger's Mouth

Bruno Touschek was born in Vienna in 1921. He was the son of Franz Touschek, an officer in the Austrian army who had fought on the Italian front during the First World War, and Camilla Weltmann, of Jewish origin. A few years after his birth, his mother contracted Spanish flu, which greatly weakened her, and she died when Bruno was ten years old.

In the 1920s, Vienna was in a troubled post-war period, the Empire that had ruled Austria for centuries having dissolved like snow in the sun. The memory of the horrors of the First World War had a long tail and would soon lead to even more serious and fatal consequences. The experience of such a brutal war inevitably had repercussions on the political life of many European countries: if the defeated ones—Germany, Austria and Russia—were dragged into revolution, the winners were instead bled and bankrupted. These were years of enormous political instability.

© Springer Nature Switzerland AG 2020
G. Mussardo, *The ABC's of Science*,
https://doi.org/10.1007/978-3-030-55169-8_20

The Versailles treaty certainly had not ensured the peace that many were hoping for. Everyone knew that it would only be a matter of time before a new disaster arose: in Italy, the 1922 march on Rome resulted in Mussolini's fascist regime; Germany followed suit in 1933, when Hitler was appointed Reich Chancellor; Austria, which, in the meantime, had become a Republic, had, after an initial democratic experience marked by immense economic difficulties, first suffered the coup d'état of Engelbert Dollfuss, and then the invasion and occupation by German troops, which eventually led to the Anschluss, the annexation to Germany, with the plebiscite of 1938.

Up to his sixteenth year, Bruno Touschek had attended high school in Vienna without any problems. In 1937, all of a sudden, he discovered what it meant to be Jewish in a country where anti-Semitism was becoming more and more virulent day by day: he was simply expelled from school. He managed, however, to get his degree at a private school and, immediately afterwards, he went on holiday to Italy to visit his grandmother and his aunt Ada, who was married to an Italian businessman. The uncle had a business that sold desalination pumps and other hydraulic systems: the family lived on the outskirts of Castelli Romani, that complex of rolling hills dotted with woods, lakes and cardinal villas, with an enchanting view that extends towards the south, in the direction of the Alban Hills. At first, Bruno had thought of studying Engineering in Rome, and had therefore started to follow Francesco Severi's course in Mathematical Analysis. Then, he realized that chemistry was more congenial to him, and he hoped to be able to study it in Manchester: the outbreak of the war, however, dissolved his dream, since it was impossible to obtain a visa for England. Hence, he could only return to Vienna. There, he began to attend some courses at the university, trying as much as possible not to be noticed. But it was all for naught: he was identified and expelled from the academic circles for racial reasons. The physicist Paul Urban, at the time, a young professor at the University of Vienna and well known for his political ideas that were contrary to the National Socialist party, had fortunately taken a liking to him and, in addition to letting him consult his well-stocked library, also put him in contact with good old Arnold Sommerfeld. By then, Touschek had already read Sommerfeld's opus magnum, the legendary *Atombau und Spektrallinien*; not only that, he had also found several small errors in the text that he had reported to the author, thus starting a correspondence with the famous German physicist. So, when Urban wrote about Touschek to Sommerfeld, begging him to help that brilliant boy find a place for his studies, Sommerfeld immediately contacted his colleague Paul Harteck in Hamburg. Although he was already retired, Sommerfeld's word still bore weight in German academic circles: Touschek arrived in Hamburg in the autumn of 1940, with the hope that no one would notice his "racial imperfection" in that city. He did not know he had, in fact, walked directly into the tiger's mouth.

The Prison Full of Bats

In Hamburg, to earn some money, he took on many jobs, some at the same time; he also attended various courses at the university, particularly those taught by Wilhelm

Lenz, Ernst Ising's mentor, and Johannes Jensen, who would receive the Nobel Prize in Physics in 1963 for his discoveries on nuclear structure. Touschek often changed houses and neighborhoods, so that he could not be easily identified. He also hoped that there in Hamburg, where nobody knew him, the Czech origin of his surname would not give rise to suspicion. One day, on the train, he met a girl who worked at a company that had changed its initial name, Lowenradio, which was too obviously Jewish, to Opta. They became friends, and the girl introduced him to the factory managers, who were happy to offer him a job. Specifically, he was hired to do the editorial work relating to one of their specialized electrical engineering magazines. It was in this way that he came across an article that would have a decisive influence on his life: in it, the Norwegian physicist Rolf Wideröe, a pioneer in the field of accelerators of elementary particles, proposed the construction of a betatron to better study the interactions in which electrons were involved. He described this accelerator machine in great detail; above all, he explained how to keep the particles in orbit once they were injected into the machine. However, there was something strange in his calculations; to Touschek, at least, they seemed incorrect, and therefore he decided to write about it to Wideröe. When, later on, Wideröe was asked by the German military command to build a betatron for unidentified war purposes, the first person he contacted was Bruno Touschek.

The project required inquiring about those electromechanical companies most qualified to build magnets, ammeters, and cathode ray tubes: Touschek therefore began to make more frequent trips to the public library and the Chamber of Commerce to consult foreign texts and magazines. At the reception desk, he asked for the new issues of the top magazines, no matter whether they were in German, French or English, as he seemed at ease with various languages. A glance was enough for him to understand what was written in those abstruse articles. He also used to wander up and down among the shelves as if in a strange euphoria, suddenly stopping, sitting down at the first available table, and writing a summary of what he had just read in his notebook. Needless to say, this behavior caught the attention of one of the librarians, who spoke of that strange boy to one of his colleagues. The colleague, in turn, requested an appointment with his superior, who reported his suspicions and his "serious concerns" at the police station on the corner of Hallerstraße: he must be a spy, he said to the officer in black uniform. Bruno Touschek was arrested by the Gestapo in January 1945.

Wideröe tried to convince the German military leaders with whom he was in contact that this was just a big misunderstanding: Touschek was absolutely not a spy. But his efforts were in vain: for the Gestapo, the arrest was more than justified by the fact that he had kept his Jewish origins secret. Wideröe only managed to gain permission to go and find him in prison, in order to bring him some food, his beloved books and his neat notebooks. When they finally met, it was a very intense afternoon, during which they immediately started talking about the stability of the betatron, the way of injecting the particles inside of the machine in an optimal way, electrons, the value of the magnetic field for their acceleration, and various other issues of physics: in other words, during that half hour of conversation, the squalor of the prison disappeared as if by magic.

At the beginning of March 1945, the order came down to transfer all political prisoners from Hamburg to a concentration camp in Kiel. When it was time to move, Touschek was in the grip of a very high fever, heavily debilitated and weighed down by the load of books he carried on his shoulders. At one point, he passed out and fell to the ground, just when the prisoners' column had reached the road and was about to get into the trucks. An SS officer yelled at him to get up, ordered him to move, kicked him in the stomach, and finally pulled out a gun and shot him, aiming at his head. The sidewalk was covered by a pool of blood, and the rest of the prisoners continued to march, trying to avoid the body lying on the ground, between the books and notebooks that were now being stained red. He was left there on the sidewalk, on that street corner, presumed dead. However, the shot had only wounded his ear, so he remained motionless, stunned by the assault and too weak to get up because of the very high fever that was devouring him. He only got to his feet after the column of guards and prisoners had left, to the general amazement of the passersby: he was staggering, but he was alive. He was rescued and taken to hospital, but there, he was arrested again following a report by the doctors. He was locked up in the prison at Altona, which he christened "the bat prison," where everything was old and decayed, including the guards. He remained there until the end of the war.

The Return to Rome

After the war, Touschek went to Göttingen, where he found the survivors of the great community of German Physics: Heisenberg and von Weizsäcker were there, both very tired by the war experience and by their long exile in England. In the Department, one could sometimes also see Pauli wandering around. In Göttingen, Touschek obtained the title of Diplomphysiker thanks to a thesis dedicated to the design and construction of a bevatron. In February 1947, he moved to Glasgow, having been granted a scholarship to continue his studies and earn a doctorate. His mentor at the time was Philip Dee, who left a vivid portrait of him:

> Bruno Touschek arrived in Glasgow and he immediately became a close collaborator of mine and a close friend. He was extremely intelligent and an original person, very outgoing and full of energy. Bruno lived his life fully, with an enthusiasm that at times consumed the energies of others. I must also say that, at that time, he gave me many problems! One was with his choice of house: he changed many, none seemed ever to suit him. One Sunday, I heard a knock on the door, and when I opened it, I found myself facing Bruno with a swollen eye and several other wounds on his face: he had an altercation with his neighbor because he had grievously insulted the latter's wife. The only solution to this house problem was to give him a room on the top floor in our house; in the evening, he used to come down to talk at a rapid pace about what he had just discovered, or to play chess, a game at which he was very good. We often went out on weekends, to walk along the paths of the Highlands, and he never failed to take a dip and a swim in the frozen lakes of those lands. He had notebooks in which there were dozens of drawings of very original subjects. Then, there were equations and formulas, all written in a beautiful calligraphy: he had great self-confidence, in fact, he consulted scientific literature very rarely; he was sure that his derivation of a result would be increasingly simple and short.

He began to return to Rome more and more often, and to visit more of Italy, thanks, in part, to his great passion for travelling on his motorcycle. In 1952, he decided to move to Rome permanently, for both cultural and familial reasons. His aunt Ada was still there and, for some time, he worked at his uncles' firm. He then began to attend the Department of Physics, where he often stopped to talk to Bruno Ferretti, who held the chair of theoretical physics: the two had a great amount of esteem for each other, shared several interests in common, and used to spend the afternoons discussing the latest news shaking the scientific world. He was subsequently noticed by Edoardo Amaldi, the only one of the famous group in via Panisperna who was still in Rome and who had taken on the responsibility of keeping the spirit of that group alive. Amaldi wasted no time in giving him a position as a researcher within the newly created National Institute of Nuclear Physics, and this proved to be an excellent decision: Touschek, in fact, quickly became one of the main protagonists of the new group of scientists gathered in Rome. In 1957, Touschek had his first undergraduate students, Francesco Calogero and Nicola Cabibbo. This was followed by a series of names of other great future Italian theoretical physicists, including Paolo Di Vecchia, Giancarlo Rossi and Luciano Pietronero. Among his colleagues was Raoul Gatto, who loved to recall how, upon the arrival of some foreign guest, Bruno would always propose to show him around in his Triumph, a convertible sports car, very fancy but extremely uncomfortable: he also drove in a reckless manner and he was fully aware of it, so much so that, before getting in, he always pronounced the Latin phrase "morituri te salutant."

His predilection for calembour and word games immediately became known in the Department: he introduced the "Bond Factor" to describe a physical exponent of 0.07 in value "in honor of James Bond"; he called his two cats Planck and Pauli, the latter obviously being "very clever"; he uttered very strange Italian phrases, stuffed with bizarre jargon, comprehensible only to a few.

The first period of his scientific activity in Rome was dedicated to a series of theoretical works on topics of great relevance at the time, for example, the chiral transformations, important for the study of the violation of the equality of weak interactions (a topic that yielded the Nobel Prize to Lee and Yang in 1957), and the temporal inversion properties of physical laws, "the reverse of time," as he called it. Incidentally, "the reverse of time" was at the origin of one of the most famous anecdotes from his life. He had bought a motorcycle, which he called Josephine, and with this motorcycle, he used to run around the hairpin turns of the Castelli Romani: he loved to tell his friends that Josephine was so smart that she knew how to bring him home safely, even after many drinks in the tavern. It was not always true, though. In fact, returning home drunk one evening, he fell disastrously from the motorbike and, when asked by the rescuers who he was and what he did, he replied that he was a specialist in the "reversal of time," leading to a consequent emergency hospitalization in the neurology ward. He insisted that it was not he who had crashed into a truck, but rather that it had been the truck that had crashed into him, as was easily understood by making a "reverse of time" transformation. Valentino von Braitenberg, the head of the hospital and a renowned psychiatrist, visited him. He concluded, in seraphic terms, "It may be that he is not crazy; more simply, he may just be a theoretical physicist."

In 1955, Bruno Touschek married Elspeth Yonge, the daughter of an Edinburgh Zoology professor, both of whom he had met while he was in Glasgow. After a few years, two children were born. The family moved to the center of Rome. Here too, finding a home proved to be a problem: the terrible experiences that he had had during the war had developed into a visceral antimilitarism, so that every time he made an appointment to rent an apartment, the first thing he would ask was whether there were any soldiers living in the building, in which case he would leave immediately. When, in 1957, the Russians sent the dog Laika into orbit on Sputnik, the comment he made to a friend was, "It is a rather expensive method, but gradually we will get rid of all the military by sending them into space."

AdA

At the Physics Department of Rome, Touschek had stimulated a hotbed of discussions, as had not been seen since the days of Fermi. These became even livelier when they began to touch on a field that he knew very well, that of particle accelerators. As a first class physicist, he had full mastery of both the experimental and the theoretical. By all accounts, he was an expert in field theory and, in particular, quantum electrodynamics, which he considered a "clean theory" that could be blindly trusted in designing accelerators, magnetic lenses and detectors, as well as in making progress in understanding fundamental interactions at the microscopic level. In the 1950s, thanks to the particle accelerators that were being built everywhere in the world—in the United States, the USSR, England and France—hundreds of new elementary particles had been discovered, so much so that the adjective "elementary" used to denote them seemed to have become an ironic oxymoron. These particles mainly interacted with the strong nuclear force of hadrons: Touschek collectively called them the "hadron thorn." In fact, contrary to the trend of the time, when all groups were studying the ravages that produced hadronic particles, Touschek thought that using electrodynamics would help more in understanding something more profound about the subatomic world.

In Frascati, the construction of a very high energy electro-synchrotron was nearing completion, destined to become the most powerful accelerator of the time. The director of the project was Giorgio Salvini, who had been working diligently on it for many years. On January 10, 1960, however, Pief Panofski came to Rome and, in the seminar that he held at the Department, illustrated the work that was being done in Stanford, California, in the group led by Burton Richter, who had built a machine to induce collisions between electrons in their center of mass. While listening to that seminar, Touschek shook his head, puzzled: he considered that experiment "stuff for students"; nothing would come out of the study of those shocks in which two electrons were simultaneously involved, due to the double initial electric charge of this state. "It would be much better," he said, "to study collisions between electrons and positrons!" The "machinists" present—as he called experimental accelerator physicists—replied that was too difficult, perhaps impossible. But, as Carlo Rubbia said, "Touschek was mentally prepared to realize the impossible, to think the unthinkable."

A week later, it was Touschek who gave a seminar at Frescati on the exciting prospects derived from the collision of matter and antimatter, electrons and positrons: the energy present at the center of a mass, in this case, would have given rise to new particles with opposite electric charges, perhaps immersed in a sea of photons, in a dance of couples produced out of the vacuum state. His calculations were reported in a beautiful notebook with a red cover; the formulas did not have a single smudge and everything seemed crystal clear. Theorists seemed very excited by these new ideas. At one point, Touschek also had the brazenness to launch an outrageous challenge, that of modifying the accelerator under construction by Salvini in order to study "all of that new physics." Salvini jumped up like a madman, screaming that he would never, ever agree to change the intended use of the machine that had cost him so many years and years of work and enormous organizational and financial effort. At that point, Giorgio Ghigo, a highly experienced experimental physicist who enjoyed everyone's respect, took the floor: he suggested that the wisest thing to do was not to change Salvini's machine, but to build a prototype of the accelerator machine that Touschek had in mind. The suggestion was accepted with enthusiasm. A formidable team of researchers soon coagulated around Touschek and his daring project: Carlo Bernardini, Gianfranco Corazza and Giorgio Ghigo himself. Thus was born the extraordinary adventure of the AdA accelerator, a name chosen by Touschek in memory of his aunt and as an acronym for the Anelli di Accumulazione (Rings of Accumulation).

AdA was the progenitor of all subsequent magnetic rings in which electrons and positrons circulate in opposite directions, and it was the first to show the fascinating research prospects for the fundamental laws of matter and antimatter opened up by colliding particles of opposite charges. The AdA accelerator was followed, also in Frascati, by the construction of Adone, after which many other similar machines were built in France, Germany, Russia, Japan and the United States. As their size increased, the energy of these accelerators also increased, and the results obtained with these new machines radically changed the course of elementary particle physics: for example, the intermediate bosons—responsible for the electroweak force—were discovered, as well as the Higgs boson, the particle by virtue of which all other particles acquire their masses.

Bruno Touschek was an unruly genius, unbridled in his drinking, but tirelessly energetic and extroverted, with a passion for physics and life that knew no limits. He was the pure embodiment of the Central European culture of Vienna at the beginning of the century, steeped in rigor, yet with flashes of humor and irony that exploded in fulminating and caricatured jokes of spirit. When we think of him, he appears sitting in a cafe, intent on discussing with other physicists the secrets of nature that AdA was discovering, amidst the smoke of a cigarette and a glass of "computer nectar," i.e. Chianti Straccali, his favorite wine.

Further Reading

E. Amaldi, *Bruno Touschek's Legacy*, CERN 81-19, Geneva (1981)

Ulam. The Art of Simulation

At sunset, the desert gave its best: the canyons with their laced peaks were slowly tinged with amaranth, the cactuses cast long shadows on the golden stones and the large fireball of the sun floated softly on the horizon, so close as to be able almost to touch it. On the desolate mesa of Los Alamos, the last violet and orange glow of dusk finally gave way to the darkness of the night.

A Desert Full of History

That desert was full of history, Stan Ulam thought, observing the spectacle from the patio of his house overlooking the ocean of sand. In the silence of the night, he seemed to hear the voices of Hans Bethe, Ernest Lawrence, Enrico Fermi, John von Neumann, and Robert Oppenheimer: together, they had changed the world. Almost all of them were dead, and at that thought, Ulam felt overcome by a great sadness. Most believed him to be a strong and confident man, affable and full of charm, with a smile that, in the old days, had made a big impression on women. In reality, it was all a fiction, one simulation among many, only better; after all, he was the wizard of simulations. Simulating joy to hide sadness, putting on that air of absolute security to conceal, from oneself and from others, a bitter truth, in this case, his intimate awareness that the end result of any human action could only be failure. The proof of this was the Manhattan Project, the project that had, for the first time, brought a handful of men into the presence of God. These scientists had shown that they could control—purely by force of logic—the most secret laws of nature, and the thought of this omnipotence had intoxicated them: the frightening fireball that had exploded on the night of July 15, 1945, from the turret of the "Trinity test" among the sand dunes of New Mexico, brighter and more powerful than any sun, had probably been the highest expression of genius ever conceived by Homo sapiens. But—Ulam considered—what remained of that exciting enterprise was now only the ghostly

© Springer Nature Switzerland AG 2020
G. Mussardo, *The ABC's of Science*,
https://doi.org/10.1007/978-3-030-55169-8_21

silence of Hiroshima and Nagasaki, both pulverized by "Fat Man" and "Little Boy", released by the B-29s at the dawn of two August mornings.

However, he loved that desert; he had loved it since his first visit back in 1943, and since then, he had been unable to detach himself from it. It was true that, looking back, the images of the various university campuses where he had taught in the past would run quickly through his mind, in Colorado, Wisconsin, California, but the focus of his images would always return to that plateau populated by scorpions, rabbits, rattlesnakes and coyotes. Far from everything, free to travel back to his native Poland in his thoughts, to remember the coffee of Lwów, the intoxicating smell of plum brandy, the discussions with his great teacher Stefan Banach and the dreams from when he was a boy, completely lost in the adventures of H. G. Wells and Jules Verne. An eternity had passed since those days.

"S. Ulam, Astronomer, Physicist and Mathematician"

Stan Ulam was born on April 13, 1909, in Lwów, Galicia, which, at the time, was part of the Austro-Hungarian Empire, into a very wealthy Jewish family. His maternal grandfather owned banks and steel factories scattered across both Poland and Hungary: in Central Europe, the name Ulam had the same importance as that of the Rothschilds in other countries. His was a carefree youth, although it was soon troubled by the rumble of cannons and the acrid smell that accompanies all wars: in 1916, the city was occupied by the Russians, and the family had to hurry to Vienna; in 1918, upon their return home, it was the turn of the Ukrainian troops. During that time of occupation, the family gave shelter to several relatives who had lost homes and belongings. For little Stan, all of that confusion was just a matter of play and fun. In addition to playing chess with his cousins, he spent the afternoons fantasizing about the stories in the pages of Jules Verne, enraptured by the narration of those adventurous journeys into the depths of the oceans or the centre of the earth. On the cover of a notebook, he wrote: "S. Ulam, astronomer, physicist and mathematician." In the family library, he had found a very interesting series of German mathematics books, including an algebra text written by the great Euler. He was struck by all of the enigmatic symbols, which seemed almost magical to him: "I wonder if one day I will ever understand them!" So it was that he threw himself headlong into an attempt to decipher that mysterious language.

As the child prodigy that he was, he quickly acquired a solid reputation, both in school and in the family circle. At twelve, he was already mastering the rudiments of relativity theory and had ventured into solving a problem, still unresolved, from number theory: that of the existence of perfect odd numbers. An integer is perfect when it is equal to the sum of all of its divisors: for example, $6 = 1 + 2 + 3$ is a perfect number, just as $28 = 1 + 2 + 4 + 7 + 14$. It is known that infinitely many perfect numbers exist, but all of them are even. However, regardless of the details of this fascinating problem, it is worth pointing out that Ulam, from an early age, had a great flair for choosing tough and intriguing research themes, with which to give free

rein to his imagination. Not only was it the aesthetic aspect of pure mathematics that attracted him—the rigorous logic of his theorems—he was also fascinated by the elegance of a discipline that was capable of articulating each step of a demonstration with an economy of means and great genius. This very aristocratic way of engaging in mathematics was, on the other hand, the hallmark of the Polish school of Lwów, with which he came into contact during his university years.

The Scottish Café

During his university period, he quickly discovered that the truly interesting problems were not those presented in the classrooms of the various courses, but those that held benches in discussions at the tables of the Scottish Café, the place where Lwów's mathematicians usually met. There, between a glass of brandy and a cup of coffee, a drop of slivovitz and the dense smoke of cigars, animated discussions took place on the lively cardinal sets of Cantor and their topology, logic and vector spaces, giving shape to ideas that would revolutionize entire fields of mathematics. The atmosphere was one of a casual, cheerful gang, disconcerted and unruly. Ulam's teacher, Stefan Banach, founder of modern functional analysis, was a chronic alcoholic, while his best friend, Stanisław Mazur, was a libertarian anarchist: the two argued endless about politics, science, theorems, absurd demonstrations and the state of affairs in Poland. Although geographically distant, Vienna's magical Central European influence was also distinctly felt in Lwów, in that Scottish Café, so refined and indolent, similar in this way to Ulam's very character.

He soon became the main instigator within that exclusive circle: Ulam enlivened the evenings with his improvised talks, his great classical erudition and his astonishing speed in formulating daring mathematical conjectures. He was only twenty when he was invited to the Zurich International Mathematics Congress. From that moment, his fame grew rapidly, and the brilliant works written in those years on the foundations of probability and set theory ended up attracting the attention of Solomón Léfschetz, the greatest topology expert of the time and director of the Department of Mathematics at Princeton; and it was thanks to Léfschetz and John von Neumann, a brilliant mathematician from Hungary who was already an influential member of the Princeton faculty, that Ulam began a fruitful overseas collaboration. For four years, he crossed the ocean several times, back and forward between Poland and the United States, until the year 1939, when, during his umpteenth return to Princeton, the outbreak of the war removed all possibility that he might return to his home country.

Box: The Spiral of Ulam
During a particularly boring academic meeting, Stan Ulam began to write the series of natural numbers on a squared sheet, in the form of a spiral. After

doing this, he began to blacken out the prime numbers, and he noticed that a rather interesting graphic structure emerged: the prime numbers surprisingly tended to arrange themselves along diagonals. Back in the office, he thought of investigating the problem on a large scale. He quickly wrote a program, and the data produced by the computer confirmed the result: thus represented, the world of numbers seemed like that of a big city, with long avenues entirely made up of sequences of prime numbers. Keeping in mind that, in the geometry of the spiral, straight lines are expressed by quadratic equations, of the form $g(n) = 4n^2 + an + b$ (with a and b integers), this exercise of "experimental mathematics" conducted by Ulam clearly demonstrated the existence of a large number of quadratic equations capable of generating sequences of prime numbers. Ulam's spiral had therefore added a touch of fantasy to speculation on the enigmatic combination of order and disorder that governs the distribution of prime numbers.

The Art of Computation

He remained in the United States from that day on: it was there that he learned of the fall of Poland, the deportation of his family to concentration camps and the looting of all of his relatives' belongings. Soon, his father's checks also stopped arriving, and he found himself, for the first time, having to deal with harsh everyday reality. Mathematics was his morphine, the most immediate way to escape from the anguish of those days. He completely immersed himself in the world of formulas, in which all emotions were dampened and worries magically disappeared: there, the power of the imagination reigned and there was room only for a logically perfect universe.

He further strengthened his friendship with von Neumann, his dear friend Johnny, who would help him in many circumstances: in addition to their common Central European roots, the two shared the same tastes in dress (both of them loved elegant clothes, scarves and soft coats), good wine, the pleasure of conversation and continuous intellectual provocation. They spent hours recalling jokes and stories about Jews, gossiping about mutual friends, poking fun at each other. At the time, John von Neumann was already an acclaimed celebrity in the exclusive academic circles of the East Coast, the acknowledged father of linear operator theory in Hilbert spaces, the rigorous formalism upon which the mathematical framework of mechanics had finally consolidated quantum. Ulam, however, had a lot of fun mocking Johnny's eager rigorousness, exposing the weaker aspects of his theories and ridiculing the more formal ones. It was a rather original way of expressing the great admiration that he had for the genius of his friend, who, rather than being offended by those pungent comments, would explode into uproarious laughter.

During that series of semi-serious conversations, however, an exciting idea was born, a project that touched on the very heart of mathematics: the art of computing.

It was a new world to be explored, but both understood that it would open up bound-less scientific horizons. It was only necessary to rely on numerical simulations and significant use of a calculator; they would leave the abstract universe of mathematics so as to be contaminated by the ways and concrete problems of physics.

The Power of the Calculator

Developing "experimental mathematics" was Ulam and von Neumann's dream, a project that seemed to express a contradiction in terms: isn't mathematics an exact science?! It certainly is, but very often, the nature of the problems that this science faces transcends the classical scheme of axioms, hypotheses and demonstrations. One need only take a look at the natural world to understand how it is an inexhaustible source of inextricable mathematical problems: even the state of the simplest matter—that of gases—of which absolutely all of the elementary physical laws are known, requires the simultaneous solution of a huge number of differential equations for its rigorous treatment, an impossible mission for anyone. Not to mention the highly non-linear and very complicated systems of equations for forecasting weather or the study of the nuclear processes underlying the life and death of a star: problems that require the processing of a very large sample of data about pressure and temperature in order to be able to bring about results that make at least some sense. But it is mathematics itself, as a good servant-mistress, that suggests problems that go beyond normal human computing skills: what is, for example, the millionth prime number? Or the first time in which the number 7 appears one hundred times in a row in the expression of the famous π? Are four colors sufficient to color all geographic maps without two neighboring countries sharing the same color?

Ulam and von Neumann understood that, to face this series of problems, it was necessary to promote the computer as a primary instrument of science. It was there-fore necessary, first of all, to develop the theory of computation ab initio, in order to overcome the thousand obstacles of electronic engineering that stood in the way of its correct implementation. The opportunity presented itself with the outbreak of the Second World War and the extraordinary war effort put in place by the United States.

Until then, electronic computers as we know them now were practically non-existent. The most sophisticated ones were slow and elephantine machines: with its two thousand tons of weight, one hundred twenty square meters of extension and half a million electric welds, the ENIAC at the University of Pennsylvania was, at the time, one of the greatest jewels of technology. Still, that monster of valves and electromechanical relays could not go beyond fairly simple and very specific calculations, such as the ballistic trajectories of interest to the Ministry of Defense. It was only thanks to the efforts of von Neumann and Ulam that, for the first time, the idea of building a calculator capable of performing any type of calculation and simulating the most varied problems, even those deemed insoluble, took hold. The first modern electronic calculator thus saw the light in the remote desert of New

Mexico. It was Ulam's friend Johnny who asked him to take part in "an unidentified physics project, which had to do with what the stars are made of." Thus, Ulam landed in Los Alamos for the first time in 1943.

While England's Alan Turing had developed the theoretical bases of computer science, with the introduction of an abstract model of programmable calculating machine (the famous Turing machine) and von Neumann had rigorously defined the concept of an electronic computer and its architecture—with the distinction between primary (ROM) and secondary (RAM) memory and a style of programming through flowcharts—Ulam instead became founder of the art of numerical simulation. The computer was the ideal machine for making predictions using the probabilistic laws of chance. He found that the key was to generate long tables of random numbers. It was a real form of arithmetic roulette and, for this reason, the procedure took on the name of the "Monte Carlo method". To estimate, for example, the famous irrational number π with the Monte Carlo method, it is sufficient to use known geometric formulas and count how many coins thrown at random into a square fall inside of the circle inscribed within it; the procedure is easily accomplished by generating a pair of random numbers (x, y) between -1 and 1 on the computer and selecting only those pairs that satisfy the condition $x2 + y2 < 1$: as the total number of attempts increases, the ratio between the number of couples that satisfy this condition and the total number of them tends gently to the number well known since high school, $\pi = 3.1415 \ldots$

Within the Manhattan Project, together with Nick Metropolis and his friend Johnny, Ulam developed sophisticated simulation methods and applied them to calculation of the statistics of neutron multiplication processes in nuclear fission processes and determination of the time of implosion of a sphere of uranium, a result necessary to create the atomic bomb. Ulam admired Enrico Fermi's incredible ability to solve complicated physics problems with the bare minimum of mathematics. Working closely with him, Hans Bethe, Richard Feynman, and J. Robert Oppenheimer—the brightest group of physicists ever gathered in a single laboratory—he began to become more and more passionate about physics problems that involved turbulence processes, the behavior of radiation, and the chain of nuclear reactions. In that desert city, known in the past only for its American Indians and the uncertain gait of their horses he ultimately built the crucial weapon that would definitively fell Japan and change the course of human history.

Looking at the desert's waning shadows, Ulam's thoughts once again went to his dear old friend Johnny, to the generous help that he had never denied him, to their endless discussions on game theory and computer laws, to those past evenings spent drinking wine and playing cards. He had been close to him until the end. He remembered the last time that he had seen him, one of their usual skirmishes breaking out despite von Neumann's weakened state; he remembered Johnny's laughter, so loud that it even shook the hospital bed where he lay. There had never been a serious word between them, but it was precisely for this reason that they shared a truly profound friendship. When Johnny died, it was the only time that Stan had cried: in that game of infinite possibilities for the world that they had imagined together, the only one that remained was an empty formula.

Further Readings

S. Ulam, *The Scottish Book: A Collection of Problems* (Los Alamos, 1957)
S. Ulam, *Adventures of a Mathematician* (University of California Press, 1991)

Venus. The Cruel Goddess

This is the story of a love, both tragic and crazy. A madness that lasted ten years, a passion that no war and no ocean was able to stop, a drama that took place between the South Seas and the coasts of India, between Mauritius and the rocks of Madagascar, between the beaches of the Philippines and the waves of the Indian ocean. An impossible love, between a glacial goddess and a very delicate and gentle man, as even indicated in his name: Le Gentil, full name Guillaume Joseph Hyacinthe Jean-Baptiste Le Gentil de la Galaiose, an astronomer at the Académie Royale des Sciences in Paris. In his attempts to attract the attention of his beloved, Guillaume lost his mind and all of his possessions; he even braved the oblivion of death for her, sailing across perilous seas and trekking through equatorial forests. The unattainable object of his desire was Venus, the goddess of love and beauty, the brightest of all of the planets.

Solar Walkways

After the Sun and the Moon, Venus is the brightest object in the night sky, visible just shortly before dawn and immediately after sunset. For the ancients, the icy beauty of this planet was so seductive that the choice of its name couldn't be anything other but Venus, the goddess of beauty. For the Babylonians, it was Ishtar, the goddess of womanhood and love. Venus is the second celestial body closest to the Sun (the first is Mercury) and the one closest to us. Although different in many respects, Mercury and Venus are twin planets. The astronomical interest in these planets is linked, in particular, to their transit along the disk of the Sun, a rather singular phenomenon. In fact, a planet that "transits" must be in an intermediate position between the Sun and the Earth, that is, its orbit around the Sun must be internal to that of the Earth. If the transits of Mercury are relatively frequent (they happen, in fact, fourteen times every century on average), those of Venus are rather rare, almost exceptional.

© Springer Nature Switzerland AG 2020
G. Mussardo, *The ABC's of Science*,
https://doi.org/10.1007/978-3-030-55169-8_22

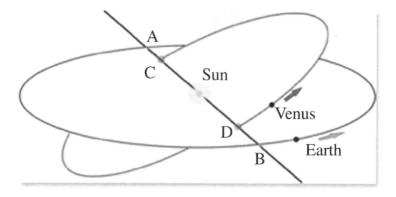

Fig. 1 Schematic view of the orbits of the Earth (blue line) and Venus (red line), together with the line of nodes: the transits can be observed when the two planets are both on the same side with respect to the Sun and simultaneously on the line of nodes, such as the points A and C in one case or B and D in the other case

Venus is a genuinely haughty diva, in the very true sense of word: royal, arrogant and impassive. Her walk on the red carpet in front of the Sun occurs only twice per century: then, the wonderful cosmic clock, with the inexorable scanning of its rhythms, does the rest. The phenomenon, in fact, repeats itself in cycles consisting of four events: if the first two steps are spaced 8 years apart, for the third one, it is necessary to wait 121 years, and then another 8 for the fourth one. After 105 years have elapsed, the cycle starts again. To appreciate how rare and extraordinary these transits are, consider that, since Galileo Galilei built the first telescope, only 8 transits of Venus have occurred (in 1631, 1639, 1761, 1769, 1874, 1882, 2004, and the last one on June 6, 2012), while the next two are scheduled for December 11, 2117 and December 8, 2125. Only a few basic notions of astronomy are necessary to understand the peculiarity of the phenomenon. The rarity of these events, in particular, is attributable to the fact that the orbits of the Earth and Venus lie on different planes. These planes intersect along a straight line, called the node line (AB), on which lies the point representing the position of the Sun: for the transit of Venus to occur in front of the Sun, the two planets must be on this line simultaneously, for example, the Earth at point A and Venus at point C, or the Earth at point B and Venus at point D. When this occurs, there is a small eclipse of the Sun; in practice, however, we only observe the passage of a tiny black dot that flows slowly in front of the solar disk, a phenomenon that typically lasts about 6 h (Fig. 1).

The Astronomical Unit

The first people to observe the transit of Venus with a telescope were the brilliant young curate Jeremiah Horrocks (he was only twenty when he predicted the event)

and his friend William Crabtree, a merchant from Manchester with whom he shared an interest in astronomy and all the wonders of the sky. It was on November 24, 1639. The two were in the small village of Much Hoole near Liverpool. Horrocks had arranged the telescope in such a way as to project the image of the Sun onto a piece of paper placed in a dark room: the image was only a few centimeters wide, but everything was ready for the big event. As soon as Venus entered the solar disk, to their eyes, the shape of the planet turned into a black drop, which, from the edge of the Sun, seemed to flow inwards. They were delighted. Beyond the beauty of this natural show, it was soon clear that this celestial walkway could be used to measure, with great precision, the Earth-Sun distance, the so-called "Astronomical Unit" (UA). An exact evaluation of this magnitude has always been of crucial importance for defining both the distance of the planets and the distance of the stars closest to us. Copernicus had said that the Sun was stationary and that the planets revolved around it; then, Johannes Kepler had established that the orbits were ellipses and determined the ratio of their major semi-axes: for example, he had calculated that the distance of Jupiter from the Sun was 5.2 times the distance that separates the Earth from the Sun, while that of Saturn was 9.5 times the distance between the Earth and the Sun. Therefore, once the distance of the Earth from the Sun was known, all of the other distances of the planets could be easily obtained.

However, there was a much more important and broader reason for trying to accurately evaluate the Earth-Sun distance: in fact, the Earth's orbit could be the starting point for measuring the distance that separates us from the stars, that is, widening our gaze from our Solar System to the cosmos. How? Just use trigonometry: it is, in fact, sufficient to measure the slight displacement of the position of a star from two opposite points of the Earth's trajectory, say, the points of the orbit relating to March 21 and December 21, and then apply a simple trigonometric formula and, indeed, in this way, easily obtain the distance of the star from the Earth. The first distance measured in this way was that of a star in the Swan constellation: this was in 1838, and the next year, it was Alpha Centauri's turn. It is therefore understandable that coming to know the Earth-Sun distance as accurately as possible became the obsession of the astronomers of the eighteenth and nineteenth centuries: the real astronomer of Greenwich, Sir George Airy—also well-known for having revealed the origin of the interference fringes present in the colors of the rainbow—once said that it was "the noblest problem in all of astronomy."

Box: The Thorny Beauty of Venus

Although the name recalls the Roman goddess of beauty, the physical conditions of this planet are far from idyllic. Venus is a planet very similar in mass and size to Earth, but its atmosphere is made up of carbon dioxide a hundred times denser than Earth's: the enormous greenhouse effect that it gives rise to makes the planet incredibly hot and totally inhospitable. There are also layers of sulfuric acid in the atmosphere that cause extremely violent electrical flashes. The thick blanket of carbon dioxide prevents heat from leaving the planet, and this is why its surface is the warmest in the entire Solar System, warmer even than that of Mercury, despite the fact that the latter is closer to the Sun. The atmospheric layer has always prevented direct observation of the surface of this planet with a ground telescope, and we had to wait for the NASA Magellan mission to find out, during the years 1990–1994, that the planet is dominated by volcanoes and huge lava fields over a large part of its surface.

The proximity of Venus to the Sun makes it difficult to see during the day. Instead, it starts to shine immediately after sunset, if one looks west, or in the early hours of dawn, turning to the east. For this reason, sailors in antiquity thought that there were two different stars, which they called the "evening star" and the "morning star", until Pythagoras came to understand that, in reality, it was a single celestial body. After the Moon, Venus is the brightest natural object in the night sky; it looks like a very bright whitish star. Venus is about 110 million km from the Sun, while "only" 38 million kilometers from Earth when the two planets are aligned with each other. A Venusian year, that is, the time that the planet takes to make a revolution around the Sun, is 224.7 days. One curious detail is its very slow rotation speed around its axis, which lasts something like 243 Earth days: this means that, on Venus, a "day" lasts more than a "year". Another interesting aspect is that this rotation occurs clockwise, unlike almost all of the other planets in the Solar System, with the other only exception being Uranus; therefore, on Venus the Sun rises in the west and sets in the east.

So, how can you get a precise estimate of the astronomical unit? The proposal had already been formulated in 1716 by the great English astronomer Edmund Halley (to be clear, the same scientist of the homonymous comet). In 1677, Halley was on the island of Saint Helena to observe the stars of the southern hemisphere, those close to the Southern Cross, but it happened that his visit coincided with one of the transits of Mercury: once he had placed the telescope on the higher tip of the island, he was able to measure the time that elapsed from the entry of the planet into the solar disk to its exit. That enlightening experience immediately allowed him to understand how to measure the famous astronomical unit: he put down his idea in black and white and published it in the journal "Philosophical Transactions of the Royal Society". The method that Halley proposed was relatively simple and was still based on the usual elementary notions of trigonometry, this time, however, accompanied by precise chronometric measurements to be performed in places on Earth that were just as far apart from each other. In fact, according to the English astronomer, it was sufficient to record the time of entry and exit of Venus from the solar disk from two different places on Earth (called contact points), and then use a simple trigonometric relationship to get directly to the calculation of the long sought astronomical unit. This idea, simple as it is, has, however, proved to be one of the most powerful tools in the study of the heavens and, since Halley, it has been used over and over again in astronomy to determine, for instance, the distances of many of our closest stars: nowadays, it is known under the name of the *parallax method*.

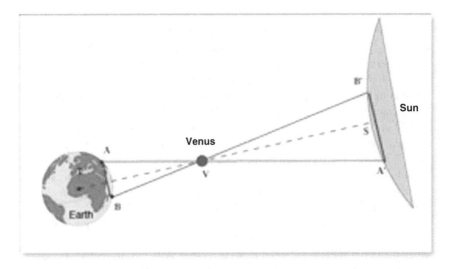

Fig. 2 The **method of parallax** is based on the displacement in the apparent position of an object viewed along two different lines of sight, and is measured by the angle or semi-angle of inclination between those two lines. In the case of Venus, two observers at A and B measure the time of transit of this planet across the solar disk. The apparent position of the planet on the disk of the sun that they measured appears to be A′ and B′, respectively. From the distance between A and B and the precise instants of time when Venus enters and exits the solar disk, it is easy to obtain the Astronomical Unit with a simple trigonometric identity.

Halley went even further in his proposal, predicting that the first useful passage would take place in 1761, and that the best places to take measurements would be Hudson Bay in Canada and a location called Bencoolen in India (Fig. 2).

The Winds of Sea and the Winds of War

Until 1760, Guillaume Le Gentil's life had been serene and happy: he got married, had several children, and had been awarded several academic prizes. He was born on September 12, 1725, in the small town of Coutances, in the western part of Normandy. After a childhood spent in those rural places, between breathtaking cliffs and rugged, wild valleys, he went to Paris to study theology at the suggestion of the prelate of the village. There, he was lucky enough to attract the interest of the French astronomer Joseph-Nicolas Delisle, professor at the Royal College. He thus turned to astronomy and, shortly after, became a close collaborator of the great Jacques Cassini, son of the most famous Gian Domenico Cassini. Le Gentil's career therefore rose very rapidly. At the age of 27, he had been appointed a member of the prestigious Académie Royale des Sciences for his studies on nebulae: he discovered the elliptical galaxy known as M32, companion of the Great Nebula of Andromeda, and the star clusters in Auriga. But all of this was not enough for him: he was an ambitious person who wanted to make an exceptional discovery that he could pass on to history and that would bring him honors, lavish earnings, the best equipment, an observatory of his own, invitations to the Court, and the praise of the king.

After the brilliant ideas and amazing inventions that had emerged in the past century—due to Copernicus, Galileo, Kepler, and Newton—the mid-eighteenth century presented itself as a relatively calm period: celestial maps of a certain accuracy had been completed, first that of John Flamsteed, then, for example, those of Pierre-François Méchain and Charles Messier; new clouds had been discovered in various constellations, such as the M78 nebula in the Orion constellation or the M79 globular cluster in the Cane Maggiore; Messier himself, who spent nights at the telescope looking for comets, was the first astronomer to spot Halley's comet in 1759; in England, William Herschel had built a telescope with a very large lens that he had ground himself and launched himself into the mapping of the celestial vault. All nice enterprises, but none really able to guarantee immortality to their authors. But the possibility of being the first to measure the Astronomical Unit would occasion a life of imperishable fame!

The trouble for Le Gentil began on March 26, 1760, when, equipped with telescopes, sextants, dials, compasses and watches, and having said goodbye to his wife and children, he embarked for Pondicherry, a city of French domination along the coast of India. The purpose of his trip was to observe the transit of Venus—scheduled for June 6, 1761—from the other side of the world, just as suggested by Halley. His ultimate goal was, of course, to measure Astronomical Unity for the first time, and thus, by right, to enter into the history of astronomy and the pantheon of science. He had left La Rochelle aboard the Berryer, a large sailing ship of the French East India Company that was headed for the Cape of Good Hope, and from there for Mauritius. The voyage across the Atlantic and Indian Oceans proceeded without any major hitches: the boat was followed along the route by herds of dolphins and small cormorants; they got supplies at the Canaries, passed near the coasts of Morocco, stopped at Cape Verde and enjoyed the suffocating equatorial climate of Douala, Cameroon for a few days. Then, they made another stop in Gabon, and then entered the tumultuous waters of the Cape of Good Hope. When they approached that promontory, the howl of the wind suddenly increased, inflating the sails and making the mast waver frighteningly: finally the slender shape of that sailboat, white like an albatross, turned its bow towards the east, pushed at twelve miles per hour by the wind that continued to blow impetuously. Once they had left the African coast behind, they set their sails towards the great Indian Ocean. They arrived in Mauritius in July 1760, where Le Gentil learned some terrible news: the war between France and England had just broken out, a war that was to be fought mainly at sea. In Port Louis, the capital of the island, the sailors were talking of nothing else: Pondicherry was besieged by the British, and therefore was unreachable! As the Berryer's commander told him personally, as far as he was concerned, their trip had ended there.

From Bad to Worse

Le Gentil was out of his mind. Every day, he wandered past the benches in the port to ask if there was any ship willing to take him to India. He also inquired about the latest news related to the conflict, asking if Pondicherry had capitulated in the meantime and what news there was from Paris. Nobody seemed to know anything at all, and everything was stopped, the large sailing ships languishing in the harbor, the military vessels just outside of the port with cannons ready to fire, the Port Louis market becoming poorer in goods. Le Gentil also fell ill with dysentery. Months of great uncertainty and trepidation followed: Le Gentil even considered the idea of changing his plans, opting for Rodrigues Island as a vantage point, perhaps one equally good. Then, he changed his mind again and rushed to the port to hear the latest news. All of this lasted until one day in March 1761, when the order arrived from Paris to send a military ship to India for reinforcements. It seemed like it couldn't be true … The vessel was the Sylphide, a sailing ship with a crew of 200 men and armed with 40 cannons. Le Gentil was assured that it was a fast ship and that, despite the contrary winds, he would arrive on the Indian coasts in time to observe the transit of his beloved Venus, scheduled for June.

Unfortunately, this did not turn out to be the case: after the first days of sailing, strong headwinds rose, the bow climbed the waves and, when it fell to the bottom, the shoulder of water made a frightening noise. The ship was pushed relentlessly towards the coasts of Arabia, and, in recovering its route, it lost a lot of precious time. When they came in sight of India, they also discovered that Pondicherry was definitely lost and it was therefore impossible to go ashore. To avoid the worst—the captivity or sinking of the sailing ship—the captain ordered that they immediately return to Mauritius, despite all of the protests, insistences and screams by Le Gentil, who wanted to disembark at any cost. They set sail again on May 30, but on June 6, the fateful day, they were still at sea: Le Gentil had to witness the transit of Venus from the ship's deck, amidst the pitching by the sea that forced him to keep to the shadows in order not to fall into the water. Nonetheless, he kept his eyes turned to the sky, fixed on that magnificent speck that walked along the solar surface. He had no possibility of making measurements of any astronomical utility: he was unable to accurately calculate the transit time, much less did he know the vessel's longitude at the exact moment, which would have allowed him to calculate the correspondent time in Paris. In a nutshell, from a scientific point of view, it was a real disaster. However, he did not lose faith: as a caring lover, he decided to wait patiently for the new passage of that spiteful goddess. After all, it was only 8 years …

The Mockery

He spent his time making cartographic and naturalistic observations on Mauritius and Madagascar, studying the plants and strange animals on those islands. It was a bizarre world of numerous endemic species, populated by ring-tailed lemurs, chameleons with more extravagant shapes and colors, bats with large wingspans, porcupines related to moles, carnivorous animals never seen in Europe, birds with very lively feathers, gigantic snakes: those places were a real sanctuary of nature. Le Gentil filled several notebooks with notes and quick sketches of the animal species to show his wife and friends once he returned to France. Five years passed quickly.

In May 1766, he decided that it was time to leave for Manila, the Philippines, which he believed could be an equally good place to observe the transit of Venus. Arriving at his destination, he immediately understood that this choice had been a big mistake. The climatic conditions were not the best and, to complicate matters, the governor of Manila was not particularly endeared of the French, or, at least, he was certainly not so of Le Gentil: he accused him of being a swindler and threatened to lock him up in prison if he didn't leave Manila soon. Thus, Le Gentil left in hurry for Pondicherry, which had returned under French control in the meantime. The navigation in the seas of China was stormy, with the ship making a necessary stop in Malacca, but he finally arrived in the Indian city in March 1768, well in advance of the appointment with his beloved.

His reception into the country went very well and, while waiting for the destined meeting, he learned the local language and studied Indian astronomy. He carefully set up the observation station, lovingly cleaned the telescopes, and calibrated the watches and all of the sextants. The weather was always splendid, the nights enchanting, but the long-awaited day, June 3, 1769, brought the most treacherous of mischief: in the morning, a cloud—one of those large, gray clouds laden with rain—peeked over the horizon, stretching out like a huge sheet over the whole sky! Venus had escaped him again, this time forever …

Le Gentil went mad with despair, wandering for a year among the forests and villages of India in a state of delirium. It was hard for him to understand the perfidious cruelty of his beloved. Back in France, news about him had dried up. However, in a rare moment of lucidity, he finally decided to go back home: after a long journey at sea, he landed on the Spanish coast and, from there, he crossed the Pyrenees, the Lower Aquitaine and the Loire valleys on foot, walking all the way through to Paris, where he arrived exhausted in December 1771. Everything seemed new and familiar to him at the same time. However, the surprises were not over yet: it had been believed that he was dead, his place at the Academy had been assigned to another astronomer, his wife had remarried, and many, even distant, relatives had inherited all of his belongings. Pirandello could not have asked for a better story for his novel "The Late Mattia Pascal". Le Gentil had to stand trial to prove to everybody that he was alive and that he was, indeed, Guillaume Joseph Hyacinthe Jean-Baptiste Le Gentil de la Galaisière, an astronomer at the Academie Royale des Sciences in Paris. His wife did not come back to him, but at least he was reassigned his position and managed to reclaim a portion of his possessions. "Women are cruel," he thought, but nevertheless, he did not hesitate to offer to Lepaute Hortense—the good-looking young astronomer at the Osservatoire who had a magnetic gaze—the beautiful flower that he brought back from India. It was a flower with small purple leaves, and is now known as Hortensia.

Further Reading

N. Lomb, *Transit of Venus, 1631 to the Present* (The Experiment, New York, 2011)

Weil. The Brahmin of Mathematics

From the narrow streets of the Latin Quarter of Paris to the chaotic and dusty arteries of Bombay, from the polar cold of the Finnish tundras to the tropical Brazilian climate, from the Alsatian boulevards of Strasbourg to the austere northeastern buildings of Princeton. A prison sentence for suspected espionage and an adventurous escape to the United States at the beginning of the Second World War. For many years, in hiding under the pseudonym of Nicolas Bourbaki, an imaginary mathematician of the equally phantom Royal Academy of Poland. No, this is not a summary of Somerset Maugham's biography, nor of a character from a spy story by Ken Follett. In fact, these are just some of the highlights from the incredible life of André Weil, one of the greatest mathematicians and most interesting and colorful characters of the twentieth century. He was responsible for ingenious conjectures on algebraic geometry, the demonstration of which involved at least two generations of mathematicians. He was responsible for the discovery of the deep links between the universe of algebraic curves and the queen of mathematics, the theory of numbers. Weil was the author of refined texts on the history of science, but also ones on lightning, and he could even write razor-sharp jokes such as that on the famous logarithmic law at the basis of academic recruitment:

> First rank scientists recruit first rank scientists, but second rank scientists tend to recruit third rank scientists, third rank scientists recruit fifth rank, and so on. If the director of the Department is genuinely interested in preserving the high quality of his Institute, he must exercise all of his power to put things in their right place, otherwise the deterioration process is destined to diverge indefinitely.

An acute flogger of academic customs and, at the same time, an enfant terrible: this was André Weil until the end of his days.

© Springer Nature Switzerland AG 2020
G. Mussardo, *The ABC's of Science*,
https://doi.org/10.1007/978-3-030-55169-8_23

In the Heart of Paris

André Weil was born in Paris in 1906, into a wealthy Jewish family. His grandfather, originally from Strasbourg, was an authoritative member of the community and was often called upon to resolve the most thorny issues that arose around him; he later moved to Paris with his wife and two children, Oscar and Bernard. The latter, André's future father, had specialized in medicine and had quickly acquired great fame for the precision and reliability of his diagnoses. Bernard married a Russian Jew, Selma Reinherz, one of the many arranged marriages that took place in those years: she had arrived in Paris after fleeing the city of Rostov following a pogrom that took place in the wake of the attack on Tsar Alexander II. However, theirs was a happy marriage, brightened not only by André's birth, but also by the arrival of his sister Simone, destined to become a famous philosopher and writer. They lived in a beautiful apartment in the heart of Paris, on boulevard Saint-Michel, a stone's throw from the Seine and the Luxembourg gardens. The window of their living room overlooked a street full of life, and little André stood for hours with his nose stuck to the glass to observe the swarming of passersby.

Those were incredible years for France and Paris. Although, in a large part of the population, the resentment over the humiliating defeat suffered years before in the Franco-Prussian war was still alive—so much so that the writer Renan had said: "France is dying, do not disturb its agony"—the reality was, fortunately, very different, and a new spirit had pervaded the country. "Nations and Empires, crowned with princes and potentates, rose majestically from all sides, wrapped in the treasures accumulated over the long years of peace. All were inserted and welded, without apparent dangers, in an immense architrave. The two powerful European systems stood opposite each other, sparkling and rumbling in their panoplies, but with a calm gaze," wrote Winston Churchill. The first decade of the twentieth century was, in fact, a period characterized by great intellectual vigor and an impressive series of successes in the field of science, engineering, painting, music, literature, philosophy and medicine. In literature, Paris, immortalized by Marcel Proust's pen, enchanted for the filigree of its characters, for the world of balls and receptions, the anarchy of its symbolist poets, the pervasive influence of Henri Bergson's thought. There was a whole crowd of expatriate writers, including Gertrude and Leo Stein, who animated the evenings of the living rooms on boulevard Saint-Germain and the elegant cafes on rue de Rivoli. During what was called the belle époque, Paris was unrivaled for its charm and the sweet way of life of its inhabitants: it had the atmosphere of a village that stretched from the boulevards full of shops of the Champs-Élysées to the woods of the Bois de Boulogne, from the windmills outside Montmartre to the student-filled premises of Montparnasse. Cabarets and can-cans were danced in bars. The roads had started to fill up with the first cars, and some sections of the Metro had already started up. Eugène Atget's black and white photographs brought to life trees, flowers, monuments, street vendors, shop windows, small markets and large buildings. Picasso had arrived in Paris for the first time in 1901, and in the capital, he had discovered both van Gogh and Toulouse-Lautrec: together with Duchamp, Apollinaire, Braque and other Cubist painters, he had made the refraction

of light aerial, causing reality to explode across a thousand layers of perspective. In the scientific field, Henri Becquerel had discovered the fluorescence of uranium salts and opened up the door to a new world that revolved around the atom, while the Curies had synthesized two new highly radioactive elements, polonium and radium, all three of them, to the increased glory of France, being awarded the Nobel Prize in Physics. The assorted inventions and advances in technology followed each other very quickly, helping to change the face of Paris day after day, thanks to the electric lighting of its streets, its cars, omnibuses, cinema halls and photography ateliers.

Every day, André and his sister Simone, accompanied by their mother, immersed themselves in that world that swarmed like a beehive: they would go out onto boulevard Saint-Michel, walking up towards rue Soufflot, passing Piazza della Sorbonne, and finally reaching the Luxembourg gardens, where they entertained themselves among the rides and the sweets stall, bombarding their mother with a thousand questions, a thousand 'why's. From an early age, André Weil had shown an extraordinary interest in both mathematics and philology, to which a great passion for travel was added in subsequent years.

His natural talent was always strongly encouraged by his mother, a woman of vigorous personality and refined Jewish culture. André was thus able to attend the best Parisian high schools. He excelled in all subjects and acquired an absolute command of various ancient and modern languages. He clearly had an edge, so much so that, at just sixteen, he was admitted as a pupil of the mathematics class at the prestigious École Normale Supérieure. Located in rue d'Ulm, in the middle of the Latin quarter and nearby the Pantheon, the École Normale was founded by Napoleon and had shaped the best of French mathematicians over the years, scientists of the caliber of Émile Borel, Gaston Darboux, Paul Painlevé and Jacques Hadamard. At the time of André Weil's entry, the latter had set up a series of seminars and conferences that quickly became very popular among the students of the École Normale, because it was characterized by an unusual freshness of spirit and themes that were at the forefront of the time.

The First Theorems

The "Hadamard seminar", known by that precise name (extraordinary, those French!), had the objective of offering to its participants a broad panorama of contemporary mathematical research through the critical analysis of the memories that its promoter received from different scientific circles in France, Germany and England. Following in the footsteps of a similar initiative set up by Carl Gustav Jacobi in Königsberg in the first half of the nineteenth century, Hadamard's seminar activity had definitively established itself in those years as a new model of work and learning. Hadamard, for his part, was well known and esteemed for having proven the theorem of prime numbers in 1896, using all of the most sophisticated tools that complex analysis offered. This important result concerned the asymptotic distribution of primes, or how many primes there are below a certain number N and how they are distributed, a problem that had, in the past, catalyzed the attention of Gauss, Chebyshev, Legendre, Dirichlet and Riemann. The latter, in particular, had ushered in the path of powerful methods of complex analysis, a direction later followed by Hadamard, but also by the Belgian mathematician Charles-Jean de la Vallée Poussin to prove asymptotically that the number of primes less than N was equal to N/ln N: in fact, the prime number theorem.

Hadamard was also an exceptional teacher, his lessons combining accuracy and dynamism, almost seeming like an adventure into unknown territories. He had the utmost consideration for mathematical rigor, but he also loved to underline the role played by intuition in the discovery and demonstration of a new theorem. He also had another talent: he could recognize the brightest students. He demonstrated it with Weil, whom he had noticed since the boy started attending his seminar: Weil was, in fact, the attendee most likely to flood the speaker with questions, accompanied by witty and rich comments, full of a thousand details. Hadamard had also discovered, with enormous satisfaction, that, in addition to his great mastery and competence, this young student also possessed extraordinary creative fertility. In fact, Weil had already demonstrated this by proving his first important theorem, called the decomposition theorem, inspired by a reading of Fermat and Riemann's original works, which allowed him to penetrate the true meaning of certain calculations made previously by Mordell on some interesting elliptical curves. Weil had decided to take this topic as the subject of his doctoral thesis and had informed Hadamard about it. The latter's reply was a surprise to him: "Weil, some of us have an excellent opinion of you. By submitting a thesis, you don't have to stop halfway. What you tell me shows that your work is very compelling, but has not yet paid off. It would be very interesting for you to complete it, to arrive at a general theorem." The thesis topic, in Hadamard's intentions, was to be the solution of the so-called Mordell conjecture, a holy grail of algebraic geometry. Weil enthusiastically threw himself into the venture, but immediately realized that the demonstration required far more sophisticated means than those available to him. He then decided to cease the work and, despite Hadamard's grumbling, presented his thesis as it was. He wrote in his memoirs: "Mine was a wise decision: I had seen things correctly, in fact, it took more than fifty years to prove that conjecture."

An Unstoppable Desire to Travel

Before writing and discussing his thesis, Weil had started traveling around Europe, a little bit out of curiosity, but also to get to know the most important mathematicians of the time. Passionate about art, he initially went to Italy to admire the Renaissance masterpieces and the vestiges of the past in places such as Rome, Naples, and Sicily. He had come into contact with both Francesco Severi and Vito Volterra, who were happy to host him for some time in their homes. The next stop on his tour was Germany, especially Göttingen, where he had the opportunity to meet Richard Courant, Emmy Noether and Luitzen Jan Brouwer, each famous in their own sector:

Courant in the field of differential equations, Noether in that of algebra and symmetries, Brouwer in the nascent field of topology. For Weil, these were extremely fruitful exchanges.

He then went to Sweden, as a guest of the mathematician Gösta Mittag-Leffler, whose home in Stockholm was a real bibliographic paradise, with rare volumes of the great mathematicians of the past and the latest issues of the various European magazines. Among other things, in 1882 Mittag-Leffler had founded his own journal, "Acta Mathematica", which had immediately established itself as one of the best in circulation; its first issues included articles by Cantor, Poincaré, and other celebrities. Mittag-Leffler was just as eclectic as Weil, and their conversations jumped from mathematics to art, from geometry to history, sometimes in French, sometimes in German, with a perilous drift towards Swedish, which Weil had not known before, but which he learned, at least rudimentarily, within a few weeks.

Weil discovered that he loved to travel by train, to immerse himself in that atmosphere in which the landscape flowed before his eyes while his mind wandered over the fascinating problems of geometry, algebra or topology. On the train, he managed to effect extraordinary concentration, and in those moments, he gained a better understanding of the nuances of the theorems that he had read for the first time in the library in Paris or of which he had previously heard in the classrooms of the École Normale. Between the steam puffs of the locomotive and the rolling of the carriages, he felt the musical sense of mathematics, the majestic immutability of his theorems, the rigorousness of his arguments. His sister was aware of all of this: he used to write detailed letters to her, with a watchmaker's precision, ranging from the beauty of mathematics to the character of Greek science, from the discovery of immeasurable numbers to the theorems of Dedekind or Gauss. In this wandering, discovering the creative process that was the basis of his mathematical work made him enormously happy. But there was something else that fascinated him terribly: it was the magical world of India, which he had approached since adolescence by reading the Bhagavad-Gita, the epic text of Hinduism, the same book written in Sanskrit that, more or less simultaneously, had also caught the attention of Robert Oppenheimer. After discussing his doctoral thesis, Weil had come into contact in Paris with Syed Ross Masood, then Minister of Education of the State of Hyderabad (to whom, among other things, EM Forster had dedicated his famous book *A Passage to India*). Before leaving to return to India, Masood asked young Weil if he was interested in teaching French culture in Indian schools. After a few months, Weil received a telegram: "Impossible to create a chair of French culture. However, a chair of Mathematics has opened up. Reply via cable if it matters." The answer was not long in coming.

Weil remained in India for two years, from 1930 to 1932, fully absorbing their habits and customs, excited by contact with a thousand-year-old culture so different from the European ones. With the same curiosity and fascination for the East that had, years before, overcome that great European traveler and intellectual by the name of Hermann Hesse, Weil immersed himself in a frantic reading of the classics of Eastern philosophy and traveled far and wide across the vast Indian continent. He also met Gandhi and participated in the most important episodes of the pacifist revolt against English colonialism. Thanks to his genius and his strength of character, he knew how to absorb all of the positive elements of this ancient civilization and incorporate them into his personality at the deepest level. He became, in short, the brahmin of mathematics. In the meantime, brilliant ideas continued to bloom in his mind; it was in India, for example, that he developed some fundamental insights into complex multi-variable functions and the generalization of Cauchy's formula.

Bourbaki, Academician from Poldevia

Back in Europe, he accepted a professorship, first at the University of Marseille, then at that of Strasbourg, thus resuming contact with friends and colleagues from the past. Many of these were, at the time, seriously dissatisfied with the conditions upon which mathematics teaching was based, specifically the use of not very rigorous and decidedly outdated books. One text under scrutiny was that of Edouard Goursat, *Cours d'analyse mathématique*, drawn up from the lessons that he had held over the years at the École Normale: all of the French mathematicians of the early twentieth century had been trained on this volume, but it was now showing its inadequacy. To remedy this situation, Weil—together with his friend and colleague Henri Cartan— proposed the creation of the Bourbaki group, destined to become a true secret society of mathematicians in the following years, with enormous influence in the field. To put things in perspective, it was, in fact, a society of a goliardic nature, for heaven's sake, but still secret: the name of its members was secret, the place and date of their meetings were secret, and the name and biography behind which they were hiding, the phantom Nicolas Bourbaki, was completely bogus. The tone and spirit of all of this initiative was really burlesque, a real pantomime: the name 'Nicolas,' for example, was proposed as a joke by Weil's wife, Eveline. Weil himself took charge of giving life to this ghost character, reconstructing his phantasmagorical biography: the name was inspired by a French general of Greek origin from the nineteenth century, Charles Bourbaki; this mathematician was an authoritative member of the Poldevia Academy of Sciences, an equally phantom institution; he was married and had a young daughter, recently betrothed. He also had several theorems to his credit, and an article even appeared in his name (this time seriously!) in the prestigious magazine "Comptes Rendus": it was written collectively by all of the members of the group as a way of decreeing its official act of birth.

To the low opinion that the young Bourbakists had towards the French scientific institutions and their manner of functioning (for example, their famous "medal war," unleashed in an attempt to eliminate the corrupt system of honors and medals set up by the Ministry of Education in 1937), they also added all of their scientific discontent. In the initial meetings, held in the Parisian cafes near the Luxembourg gardens, a set of dictates for regulating the life of the group was put down and a great program of rewriting and axiomatising many branches of pure mathematics was established, from analysis to algebra, from number theory to topology. Among the strict rules to be respected, there was one declaring that membership in the Bourbaki group ended at the age of fifty. Another rule concerned the genesis and management of each plan of the work that they wanted to compose: there must always be an initial collegial discussion of the topics to be treated, in which one member of the group would act as a scribe, one or two more members would work on the draft, and finally, the serious editorial work would be carried out by a third component, a task that, for a long time, was relegated to Jean Dieudonné, because of the style of his mathematical writing, which was precise and fast.

In the sixty-five years since the birth of the group, within which many eminent mathematicians (almost all French) have taken their turns, the program has manifested itself in ten full-bodied volumes that have changed the course of mathematics. To tell the truth, there has always been a fierce criticism of this pharaonic project. Many, in fact, have always cordially detested Bourbaki, both for the style of his texts and for his total inattention to the fields of applied mathematics and logic. But, despite these indisputable omissions and limitations, the opus magnum of the over 7000 pages of Bourbaki's *Éléments de mathématique* remains, to this day, a legendary monument of mathematics and a point of reference for all scholars.

The Queen of Math

"Mathematics is the queen of science and number theory is the queen of mathematics," said Gauss. The aim of this fascinating branch has always been mainly to study the properties of integer and rational numbers beyond the familiar manipulations of daily arithmetic. Since ancient times, some simple and, at the same time, curious relationships have been discovered, such as the identity $3^2 + 4^2 = 5^2$, linked to the possibility of building a right-angled triangle with sides equal to 3, 4 and 5. Other right-angled triangles with sides expressed by integers had also been identified, and had led to infinite triads of integers a, b, c linked together thanks to the relation $a^2 + b^2 = c^2$. The problem of seeing whether such a relationship also applied to powers greater than 2 had intrigued Pierre de Fermat in the seventeenth century, and the equation $a^n + b^n = c^n$ took more than 300 years before it was resolved.

In fact, number theory has the singular peculiarity of being a field in which the questions are simple, but the answers can be damned complicated. For example, the ancient Greeks were interested in the so-called "perfect numbers", those numbers equal to the sum of their divisors. Nowadays, we know the general formula of perfect even numbers, but so far, we don't know any odd ones, and we don't even know if such a thing exists. Modern arithmetic has also largely benefited from the notion of prime numbers and the incredibly elegant proof given by Eratosthenes of their infinite number. The solution of algebraic equations exclusively in terms of rational numbers was the problem that attracted the attention of Diophantus in the third century B.C., and since then, it has greatly influenced the development of this discipline. Associated with the Cabal until the Renaissance, and later considered a somewhat esoteric branch of algebra or analysis, number theory acquired its authority and autonomy only in the early eighteenth century, mainly thanks to Euler, and then, subsequently, the great mathematicians of the nineteenth century, Gauss, Lagrange and Legendre.

André Weil was a passionate lover of number theory, to which he dedicated himself energetically throughout his life. It was a subject that went very well with his style: the almost monastic rigor of ideas, combined with a rare breadth of theorization [?], typical of the best mathematicians. Weil's demonstrations were usually characterized by an economy of means and an extraordinary lucidity of presentation, almost like a sculpture by Michelangelo, whose characters lived inside of the stone until he freed them with the chisel. Weil's most significant contribution is linked to a variant of the famous conjecture on the zeroes of the Riemann function, a hypothesis that is still one of the most important open problems of mathematics today. If we may say so, the place where this theorem was born was quite unusual: like Marco Polo's *The Million*, it was, in fact, written in a squalid cell within a prison. The story is unbelievable and, as Weil himself wrote in his biography, it was "a tragicomic work in five acts: Finnish escape, arctic interlude, guardian under lock and key, serving the homeland, a farewell to arms."

At the outbreak of the Second World War, Weil was, in fact, a tourist in Finland, where he decided to stay to avoid his military obligations. But, with the onset of hostilities between Russia and Finland, local police, convinced that he was a spy, arrested him, exhibiting a series of Russian letters in his possession as proof of his guilt. For the record, they were letters dedicated to geometric themes, exchanged with the Russian mathematician Pontryagin! Having escaped from a prompt execution thanks to the fortuitous intervention of an eminent Finnish mathematician, André Weil was eventually repatriated to France, where, before his trial for desertion, he spent several months in the prison at Rouen. His strength of mind did not abandon him, even under such dramatic circumstances: in fact, the above-mentioned theorem and a series of letters on major topics of philosophy and politics addressed to his wife Eva and his sister Simone belong to that period. His sentence turned out to be a forced enlistment in the French army, and his battalion was sent shortly thereafter to England. And there, taking advantage of the general confusion, he finally managed to embark on a ship bound for the United States, later to move to Brazil. He returned permanently to the United States in 1957, first to Chicago and then, in 1958, to Princeton, where he remained until the last of his days.

At Princeton, he met Gödel, the famous Austrian logician. Gödel's famous theorem, which shook mathematics to its roots, was summarized by Weil as follows: "God exists because mathematics is consistent, but the Devil also exists because we cannot prove it!".

Further Readings

A. Weil, *The Apprenticeship of a Mathematician* (Birkhäuser, 1991)

A. Weil, *Number Theory, an Approach Through History from Hammurabi to Legendre.* Modern Birkhäuser Classic Series (2007)

X-Rays. Seeing the Invisible

For Christmas of 1895, in addition to the usual greeting card, physicists from all over Europe received an envelope containing the X-ray of a long-fingered hand, adorned with a large ring: it was the hand of Mrs. Röntgen, portrayed in the world's first X-ray images, taken by her husband, Wilhelm Conrad Röntgen, professor of physics at the University of Würzburg, Germany. The envelope was accompanied by a brief ten-page memory, in which the author explained the discovery that he made in the field of "cathode rays". The news instantly traveled around the world, and immediately consecrated the name of its author in the eyes of the general public.

If someone were to tell us today that there is a form of radiation—that of neutrinos—so penetrating that it crosses the entire sphere of the Earth without being absorbed, it is likely that we would not be so surprised. But in an era when only light was known and the study of infrared and ultraviolet radiation had barely begun, the discovery of a radiation capable of passing through the entire human body and making it seem transparent, well, this was something truly extraordinary, capable of hitting the imagination in a big way.

Seeing the invisible—what a wonder! As often happens when an important discovery passes too quickly into the public domain, there were those who misunderstood the possibilities that it offered. There were even those who came to consider X-rays dangerous for female modesty: a law in New Jersey in the United States forbade the use of X-rays in the binoculars of theatergoers—in fear that they could serve to see through clothes—while a London firm immediately launched a collection of X-ray proof underwear onto the market. Using his own words, Röntgen had unleashed "an uproar," capable, however, of opening up a new path for understanding the mysterious laws of the subatomic world.

The Mystery of the Tubes

Faraday had been the first to get excited about the strange effects of electricity in liquids: he had dissolved silver nitrate in a container of water, placing two metal

© Springer Nature Switzerland AG 2020
G. Mussardo, *The ABC's of Science*,
https://doi.org/10.1007/978-3-030-55169-8_24

plates at its ends that he then attached to a battery. He had subsequently observed a strange stirring in the water, with bubbles that seemed to walk from one plate to another. Looking more closely at the plate attached to the negative pole of the battery (the cathode), he noticed that a layer of silver had deposited there. In this way, he had drawn further confirmation of the atomic hypothesis of matter, a concept that was still in the rudimentary state at the time, but that can serve as a guide to the experimental measurements and infuse them with some sense. As usual, Faraday had scrupulously taken note of what he had observed and spent the following days making everything as quantitative as possible. In liquids, it was clear that a potential difference caused both a transport of electric charge and one of matter—on this point, there was no doubt: the salt dissolved in water gave rise to electrically charged molecules, so-called ions, which, in moving, gave rise to electricity and mass transportation.

Towards the end of the nineteenth century, the cathode ray tube became the main research tool in physics. In the original intentions of its designer, Heinrich Geissler, it could be used to study the effects of electric discharges on rarefied gases, but, in reality, this instrument soon ended up anticipating the role of the cyclotron, both as an accelerator of subatomic particles and as a forge for Nobel Prizes. Geissler had made a name for himself in the German physical community since the first half of the nineteenth century, demonstrating exceptional skills in working with glass and building mechanical devices. He had golden hands and a keen mind: thermometers, hygrometers, altimeters and extremely accurate vaporimeters were all produced in his laboratory. When he built his first tube, many were enchanted. Despite its simplicity, this experimental apparatus incorporated the most advanced technology of the time: it consisted of a glass tube about thirty centimeters long, at the ends of which were placed two metal plates; everything was put in communication with a vacuum pump, an element of crucial importance for the experiments that were intended to be carried out.

By connecting the two metal plates to a high voltage generator, one could witness a very interesting phenomenon: along the tube, colored streaks became interspersed with dark areas, and a ghostly light began to emanate from inside. By decreasing the gas pressure, the light coming from inside the tube decreased in intensity until it completely disappeared, giving way to a luminescence radiated solely from the walls of the tube. Thanks to the considerable improvements made in regard to vacuum pumps in the last decades of the nineteenth century, one could be sure that only very few gas molecules remained inside of the tube. Therefore, excluding that these were the cause of the residual luminescence, the problem arose of understanding what the origin of the new mysterious phenomenon was. Soon, it was discovered that it was the cathode that emitted such radiation, but opinions were discordant as to the nature of these so-called "cathode rays." Were they waves or particles? Opinions changed according to nationality: Heinrich Hertz, and all other German scientists in chorus with him, leaned towards the hypothesis of a new species of radiation, while William Crookes and all British scientists argued that they were electrically charged particles. Whatever they were, however, it was commonly agreed that they traveled in a straight line, as Crookes himself showed in a spectacular version of the experiment: after changing the shape of the tube, he mounted a Maltese cross made

of paraffin inside of it and placed it along the trajectory of the rays, observing that the glass showed no fluorescence where the cross prevented the cathode rays from passing through. It was not clear, however, whether these rays could be deflected by magnetic fields: both for the inexperience of the researchers of the time and for a particular experimental reason that was understood only later, the first experiments gave somewhat contradictory results in this regard, thus leaving the nature of cathode rays in a state of mystery for a few more years.

The Discovery of the Electron

All of the experiments on cathode rays were carried out in a dark environment, in order to better analyze the weak light produced by the tubes: the scientists scrutinized these rays with all possible attention in the hope of catching some details that would allow them to understand the subatomic secrets that the rays hid. The dark environment of the laboratories and that strange ghostly light created a scene more out of a sorcerer's practice of their magic than of scientists: after all, Crookes himself believed in spiritualism, and he performed experiments to deepen his studies on the supernatural; he said that, with cathode rays, he had managed to touch the boundary that separates matter from the forces that shape it, "a shaded area between the known and the unknown that has always exerted on me a particular seduction, since it is the seat of the final reality, a subtle, fruitful, magnificent reality.".

The decisive steps towards solving the puzzle were first taken by the Frenchman Jean Perrin, and then by the Englishman Joseph John Thomson. The former was able to demonstrate that the cathode rays could actually be deflected by a magnetic field and captured in a Faraday cage, while the latter finally effected their long-awaited identification: the cathode rays were nothing more than electrons!

In a famous experiment performed at the Cavendish Laboratory in Cambridge, Thomson was, in fact, able to measure the charge/mass ratio of an electron, noting that this figure was perfectly in line with other measurements made on the electrons of alkaline atoms by Pieter Zeeman in Holland, and that this ratio was about 2000 times smaller than that of the lightest ion, that of hydrogen. A very important result thus emerged: electrons of different origins behaved in the same way, and therefore they were surely a universal constituent of matter. All of this took place in the early months of 1897, but, at least in the beginning, the novelty of these remarkable discoveries was obscured by the news that had been in the international newspapers for some months: the discovery of X-rays by Röntgen.

Box: X-Rays, Relatives of Light

X-rays are a particular form of electromagnetic radiation, analogous to light, but with a significantly shorter wavelength: that of X-rays is about 3000 times shorter than the average wavelength of visible light. Their whole range is between one millionth and one billionth of a centimeter, and therefore they are extremely energetic (in fact, the energy is inversely proportional to the wavelength). But how does this penetrating radiation originate? If we send a cathode ray, formed by very fast electrons, towards a glass or metal plate, when it comes into contact with the target, these electrons are abruptly stopped: the "projectiles" thus completely lose their kinetic energy, which is, however, transformed into electromagnetic energy, in the form of X-rays.

The proof of the electromagnetic nature of X-rays was obtained in 1912, mainly by Max von Laue. The greatest difficulty was related to their very small wavelength; this is usually measured using so-called dispersion grids, plates marked by thin linear incisions (the size of the wavelength that you want to measure), located a short distance from each other. If a ray of light affects the plates under a particular angle, interference fringes are produced and the wavelength of the light can be deduced from them. However, the method was not applicable to X-rays, because, beyond a certain limit, it is difficult to increase the precision of the incisions and their spacing. Reflecting on these difficulties, Max von Laue remembered that the internal structure of crystals could provide reticles of the desired size. It wouldn't be completely safe, because, at that time, the reticular nature of the crystalline structure was itself the object of hypothesis and had not yet received experimental confirmation. But the first experiments quickly dispelled the doubts: by sending the X-rays through some crystals, these revealed the wonderful arrangement of the atoms in the crystalline structures and, moreover, made it possible to measure the wavelength of the X-rays! Following these important developments, Max von Laue won the Nobel Prize in Physics in 1914. Since then, X-rays have been used to penetrate deeper into the structure of the atom, to observe that radiation can act both as a wave and as an elementary particle, and to discover that it carries energy in defined and separate quantities, like many small pieces: the quanta.

An Important Bearded Man

William Conrad Röntgen was an imposing-looking man with a calm and deep voice. He wore a patriarchal beard that gave him pride and austerity. He was born on March 27, 1845, in Lennep, in the Rhineland; his father was a merchant and fabric-maker who had married a cousin, Charlotte Frowein, originally from Amsterdam. Wilhem spent his youth in Holland, attending Utrecht high school. Shortly before his final high school exam, however, an unpleasant episode occurred that led to the sudden interruption of his school career: having refused to identify the individual responsible for the caricature of a professor, he was expelled from the school. Without a diploma, however, he was unable to enter university, at least in Germany. He then enrolled in the Faculty of Engineering of Zurich Polytechnic, from which he graduated in 1866. However, the lessons of the famous German physicist Clausius, one of the founding fathers of thermodynamics, had aroused his interest in physics, in particular, the study of gases.

Röntgen entered the world of science in 1870, as an assistant to the physicist August Kundt, making his debut with a work on the specific heat of air, which first earned him a chair at Hohenheim high school and, subsequently, at the University of Strasbourg and Giessen. In 1888, he agreed to go to Würzburg, a good, but certainly not prestigious, university. In 1895, he was fifty years old, having had a perfectly respectable, though not extraordinary, scientific career and having gained a solid academic reputation, testified to by his appointment as university rector. Up to then, he had written 48 papers. But it was the 49th that made a bang.

That Damned Radiation from that Damned Tube

In early November 1895, Röntgen decided to repeat some experiments on cathode rays. He mounted a series of Geissler tubes, calibrated them, and began making his initial observations. As it turns out, November 8 is the date of birth of X-rays: in the evening, while everyone else had already left the laboratory, he observed that a special light was emanating from one of the tubes that he had wrapped with black paper, such as he had previously done with various grades of papers to determine the degree of transparency. For one of those imponderable reasons of chance, an open box containing photosensitive powder was located not far from Geissler's tube. And here, for no apparent reason, that dust had begun to emit light!

Warmly surprised, he grabbed the first object that came into his hands, a wooden lath, and found that, by interposing it to the presumed radiation of the tube, the luminescence of the powder diminished, but did not disappear completely. The scientist then thought to find out what would happen if he placed a screen of barium platinocyanide, a fluorescent substance that was not lacking in any physics laboratory at the time, in front of Geissler's tube: he discovered that the screen immediately began to shine. He also noted that if the tube current was interrupted, the fluorescence would disappear. The rays that produced it therefore undoubtedly came from that tube, but were absolutely invisible. However, for Röntgen, the most surprising occurrence of that evening was that the fluorescence did not cease even if a book of a thousand pages was placed between the tube and the luminous screen. His conclusion was therefore that the radiation possessed an enormous penetration capacity. Röntgen did not remember any other equally powerful radiation: he was therefore faced with a real mystery. For this reason, he decided to call them "X-rays".

In the following days, Mrs. Röntgen noticed an extraordinary excitement in her husband. He seemed to be grappling with something really important that he absolutely did not want to talk about, and he was so passionate about the matter that he had his meals and bed brought to the laboratory: in those days, for Röntgen, there was nothing but that—the damned radiation that came from that damned tube. In fact, he had discovered another incredible thing: while handling the lit Geissler tube, one of his hands had come between the tube and the screen and, in that fraction of a second, Röntgen had been able to see the shadow of his bones projected onto the screen! After repeating that experiment several dozen times, he had to conclude that, with the mysterious radiation, it was possible to see inside of the human body. On December 22, 1895, he produced the first real X-ray of a hand, that of his wife, showing the perfect definition of her bone system. It was then that Röntgen exclaimed, "And now the storm can break out!" A few days later, on December 28, he presented his first scientific report on this phenomenon, entitled *On a new species of rays*, to the President of the Physics Society. The President ordered that the work be printed immediately, and the author distributed some extracts to colleagues, sending them together with greetings for the Christmas holidays and a copy of the X-ray.

One of these letters was received, in Vienna, by a certain Flexd who showed it to his friend, Professor Lecher of Prague, who had just arrived that day to visit his father, the chief editor of a great Viennese newspaper. That same night, father and son wrote a short article for the newspaper about the sensational discovery by Röntgen: in the printer's haste, he was dubbed "Mr. Rautger of Vienna." So, the first announcement of the discovery came from Vienna, not from Würzburg, and the day after, the London press took up the news and spread it everywhere by telegraph; by January 8, 1896, it was in print throughout the American press as well. Finally, on January 9, the news also appeared in a newspaper in Würzburg, after having practically traveled around the world. But, in the city, nobody could have imagined that "Mr. Rautger of Vienna" was none other than their fellow citizen Wilhelm Conrad Röntgen.

The First Nobel Prize in History

At that point, all of the laboratories in the world, having some form of Geissler tube and a current generator, proceeded to do research on X-rays. Röntgen received innumerable letters of invitation, and the Kaiser himself called him to Berlin to decorate him with a high chivalric honor. His discovery triggered a large number of applications in science and technology. It was immediately understood that X-rays could be very useful in medical investigation: this was more than evident, but the first to use the discovery were mainly young and penniless doctors, who had to face the skepticism and incomprehension of their larger luminaries, thus often finding themselves forced to work in hospital basements with equipment that they paid for out of their own pockets. But it was also understood that the new radiation hid an atrocious pitfall: without adequate protection, it produced permanent damage to the tissues exposed to it, and many doctors discounted the audacity of its use because of the serious mutilations and even death caused by leukemia that could result. In the meantime, Röntgen enjoyed the deserved honors: he received an honorary degree from the Faculty of Medicine of the University of Würzburg and, shortly thereafter, he was awarded honorary citizenship.

In 1900, he accepted the offer of a professorship at the University of Monaco, and with it, the direction of the Institute of Physics. Despite all of this celebrity, he was always a modest man, and never thought about the social and economic advantages that the discovery could bring him. To the representatives of a famous German industry who had proposed that he file patents, Röntgen replied dryly, "Patents? I don't want any, it's not me who invented the X-rays, they belong to those who use them. Moreover, I think my discoveries have been made for humanity and should not be reserved through patents for individual companies." In 1901, he was the first physicist to be awarded the Nobel Prize.

Further Readings

H. Kragh, *Quantum Generations* (Princeton University Press, Princeton, 2002)
M.H. Shamos, *Great Experiments in Physics* (Dover, New York, 1959)

Yang-Lee. The Puzzle of the Mirror

Chinese Fairy Tales

According to a Chinese legend, reported by Jorge Luis Borges, there was a time when the world of men and the world of mirrors were two separate realities; between them, there was no coincidence of shapes or colors, and each had its own laws and customs. Peace reigned between the two worlds, and one could enter and exit the mirrors without any difficulty. Until, on one bad day, the people of the mirror decided to invade our world: they presented themselves on the border with a large army, and the war was inevitable. At first, many cruel battles took place, bloody clashes, with many deaths. In the end, however, the magical arts of the Chinese emperor managed to defeat the invaders, killing a large multitude of them and capturing all of the others. The surviving adversaries, it was decided, would be imprisoned in the mirrors and, as punishment, they would have to repeat all of the acts of men, one by one, without neglecting any; every gesture of the limbs, every movement of the face had to be perfectly duplicated. Reduced to slaves, the inhabitants of the world of mirrors were also deprived of their strength and their physical existence. It is, however, believed that this situation, which still continues to this day, will sooner or later cease to be: there will be a moment, in fact, in which the inhabitants of the mirrors will wake up from their torpor and want to return to their original lives, freeing themselves once and for all from the yoke of imitating our every gesture. According to legend, the inhabitants of the mirrors will then find allies in the inhabitants of the oceans, and the sound of weapons will be heard rising from the bottom of the mirrors.

A Zoo of Particles

Since the time of the Greeks, humankind has wondered about the composition of the world, about what the elementary bricks may be that, when properly assembled, give rise to the enormous variety of things in our world. To such a fascinating question,

© Springer Nature Switzerland AG 2020
G. Mussardo, *The ABC's of Science*,
https://doi.org/10.1007/978-3-030-55169-8_25

there have been, in the past, intriguing, though sometimes singular, answers. At first, Hellenistic philosophers, for example, through that the whole world could be interpreted as a composition of four types of substance: earth, air, fire and water. Others restricted the field to water and fire alone, still others to air and earth. More refined thinkers, such as Leucippus and Democritus, instead conceived the idea that every fragment of matter was made up of a multitude of infinitesimal elements, tiny particles capable of aggregating and melting, hovering in space or recomposing themselves. They called these "atoms," which was synonymous with "indivisible."

It took about two thousand years, and the discoveries of countless scientists— Lavoisier, Avogadro, Mendeleev, Davy, Ramsay, and many other chemists along with them—to finally gain authoritative confirmation of this atomistic vision of matter, the grammar of which can be found in the iconic image par excellence, the famous periodic table of the elements, with its more than 90 different elements, well grouped according to their chemical affinities.

At the beginning of the twentieth century, the atom was considered the last stage of matter, the indivisible unit par excellence. But, in 1911, Ernest Rutherford, with his famous experiment on gold foils bombarded with alpha particles, showed that things were much more interesting than previously thought, because he discovered that, within each atom, there was actually a much smaller unit, the nucleus: the latter had a positive electric charge and around it rotated, like so many planets around the Sun, the electrons, whose charge was instead negative. And, just like in the Solar System, in the case of the atom, the mass was concentrated almost entirely in the nucleus. In light of this great discovery, it therefore became illusory to think that the ultimate constituent of matter was the atom. It soon became clear that even the nucleus did not represent the final entities of matter. In fact, it was discovered that the nucleus, in turn, was composed of protons and neutrons, particles of practically the same mass, but different in electric charge: the proton had a charge equal and opposite to that of an electron, while the neutron, as the name itself says, was neutral. At the dawn of the 1930s, the situation thus looked like this: to each chemical element corresponds an atom; this is made up of a nucleus, formed by positively charged particles, protons, and electrically neutral particles, neutrons; around this nucleus rotate negatively charged particles, electrons; the number of protons, different from element to element, equals that of the electrons and determines the chemical properties of the various elements.

But even this brilliant synthesis was destined to collide with a much richer and more composite reality: it was seen, for example, that the neutron, left alone, could disintegrate, giving rise in the process to a proton, an electron and a new elusive particle with mysterious properties: the neutrino. Even this, however, was nothing compared to what we began to discover by studying the composition of cosmic rays or when, immediately after the Second World War, the first particle accelerators were built, huge machines with bombastic names such as "Betatron," "Electrosyn-chrotron," or the like: thanks to powerful magnetic fields and intense radio frequency pulses, these machines were able to accelerate protons or electrons and make them collide with each other with great energy. The result of these collisions was a swarm of new particles, a real zoo that had quickly started to populate the nightmares

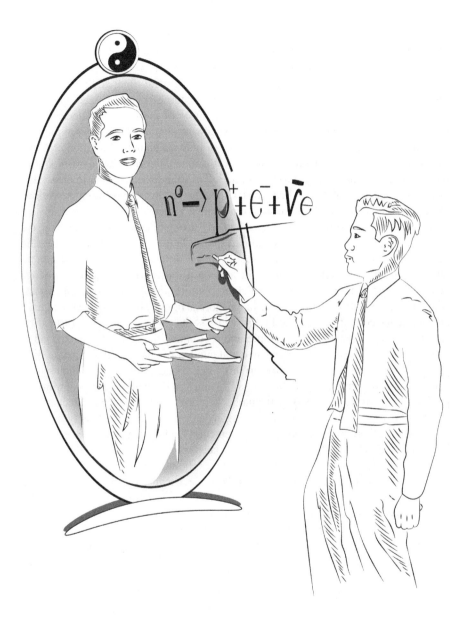

of theoretical physicists, giving them the impression of having to untangle a jungle. As early as the mid-1950s, over a hundred particles were known, and physicists had long been forced to use the letters of the Greek alphabet to give them a name.

The world of elementary particles had become extremely crowded: it was necessary to find a way to bring order to all of that chaos, that is, to understand whether or not there was a logic in that proliferation of new infinitesimal pieces of matter that responded to the names of muons, mesons, pions, hadrons, and leptons; these particles were created in collisions, lived for an ephemeral time, and then decayed into other particles of ever-lower mass, in a cascade of nuclear reactions whose traces remained etched into the photographic plates of the gigantic detectors. The first important thing that came to be established was that, in spite of these hundreds of elementary particles (a designation that, in light of their proliferation, seemed truly comical), in Nature, there are only four fundamental forces through which they interact with each other. These forces are mainly distinguished by their intensity and their range of action: all things being equal, the relative intensity between the various forces gives us very important information on their effects.

The most intense force there is the so-called "strong interaction", which acts, for example, on protons and neutrons and is responsible for the stability of the atomic nucleus. Its range of action is very small, on the order of the size of an atomic nucleus, about 10^{-13} cm; in general, it acts on a much wider class of particles, generally referred to as "hadrons", which are, in turn, divided into two subclasses, whose characterization, using modern language, can be done by following their quark and antiquark compositions. The first subclass is that of the "baryons", formed by an odd number of quarks (for example, the proton and the neutron); the second is that of "mesons", formed instead by an equal number of quarks and antiquarks (like pions). Not all particles are sensitive to this force: in fact, the electron, the relative neutrino and the muons of higher mass, together with their relative neutrinos— particles collectively indicated by the name "leptons" —are free from direct effects that result from the strong interaction. Going in decreasing order of intensity, after the strongest one, there is "electromagnetic interaction", which acts on all particles that have an electric charge. This is the force that keeps the electrons tied to the nucleus, the atoms to the molecules, the molecules to the liquids and solids: it is therefore the force that most directly influences our daily life, because it is responsible for the chemical properties of all of the elements. Compared to the strongest interaction, the electromagnetic one has an intensity approximately 1000 times smaller, but, on the other hand, it is long-range (a property that is due to the null mass of the photon, the elementary particle of which light is composed). Once, it was thought that electricity and magnetism were two distinct forces; it was James Clerk Maxwell who came to understand that, in reality, they were only two sides of the same coin, thus realizing their unification. A logical consequence of this great theoretical synthesis was the prediction of the existence of an electromagnetic radiation that could present itself in the form of visible light, radio waves or X-rays.

Continuing downward in intensity, there is the "weak interaction", which is obviously more feeble, about 10^{-13} smaller than the strongest one; however, it is responsible for most of the decay of elementary particles. The discovery of this force is credited to Enrico Fermi, who, in 1933, in order to explain the beta decay of atomic nuclei, hypothesized the existence of an interaction that simultaneously involved four fermionic particles, thus opening up the possibility that they were mutual transmutations. It is, in fact, thanks to the weak interaction that, for example, a neutron transforms into a proton, with the consequent emission of an electron and a neutrino. Yes, neutrinos … these enigmatic particles are the only ones that are sensitive solely to the weak interaction. Finally, to close the circle, there is the "gravitational interaction", the one that determines the attraction between the planets and between all bodies that possess mass; it was the first to be discovered and studied, thanks to Newton's law of gravitation, and was then reformulated in terms of space-time geometry by Einstein. Although, at the astronomical and cosmological levels, it is the most relevant force, fortunately, it plays a marginal role at the level of elementary particles, and therefore, in the first approximation, we can neglect it, thus simplifying our lives a little bit.

Two Chinese in America

The first photo that captures them together in America is a black and white photo from 1947, taken on the campus of the University of Chicago, in front of the hood of a Hudson Commodore: two young friends, almost kids, with big beautiful smiles printed on their faces, one of them very skinny, the other a little more fleshy, both with an air of happiness as people who are about to leave for a nice trip can have. And, in fact, immediately after that shot was taken, they drove by car to California, on a tour that led them to cross the whole Midwest, stopping to visit the Grand Canyon, Yosemite and the other wonderful national parks with breathtaking views, finally reaching the California coast with their arrival in San Francisco. It was an unforgettable journey, although it ultimately made them feel even more love for the land from which they had come, distant China. The protagonists were Chen-Ning Yang and Tsung-Dao Lee.

Chen-Ning Yang, born in 1922, in the Chinese province of Hefei, so frail as to seem like it was a miracle that he could even stand, had proven to be a very curious and lively person from an early age. His father was a famous local mathematics professor, who was very involved in his early education, immediately instructing his son about the fact that world is made up of numbers, equations, geometry. Once he grew up, Yang enrolled at Southwest Associated University in the city of Kunming, where he successfully achieved a master's degree in physics. Professor Wu, who was his supervisor during his thesis, had introduced him to atomic spectroscopy through the laws of symmetry and group theory, a branch of mathematics that, in those years, was beginning to attract the attention of many theoretical physicists in different parts of the world. To further increase his curiosity and, at the same time, his competence in the field of symmetries, he was assigned the book *Modern Algebraic Theories*, written by L. E. Dickinson, a volume that had fortunately been passed down to him by his father. During his university years, he had been able to read the famous articles by Albert Einstein, Paul Dirac and Enrico Fermi, and he was so impressed that he decided that he would become a physicist.

Mathematics and physics were not his only interests, however: he had enthusiastically devoured Benjamin Franklin's autobiography and had developed great admiration for this multifaceted American scientist and politician. When, following a scholarship, he arrived in the United States towards the end of 1945, he decided to call himself "Franklin", or, rather, "Frank": this immediately caught on with his friends and the other students at the University of Chicago, in which he was enrolled, and, from then on, he was universally known as Frank Yang. In the United States, he started working as an experimental physicist, but with far from exciting results: he could never calibrate the instruments well, something inevitably went wrong, he always broke some apparatus and the other students mocked him heavily: "Where there's a bang, there's Yang!" He decided that laboratory life was not for him and moved into the arms of theoretical physics. Under Edward Teller's supervision, he quickly finished his doctoral thesis on the theoretical analysis of a series of nuclear reactions and, in 1949, with a postdoc position, he moved to Princeton, to the Institute for Advanced Study, the elite home of theoretical physics: there, walking in the corridors, you could meet Albert Einstein or Wolfgang Pauli, Shinichiro Tomonaga or Robert Oppenheimer, who was the director at the time.

Tsung-Dao Lee, four years younger than Yang, was born in Shanghai in 1926, the third child of an entrepreneur. He completed his high school studies brilliantly in Kanchow, in the province of Kiangsi, but, following the invasion of China by the Japanese, he was forced to move to Kunming, where he enrolled in the Southwest Associated University. It was there that the two met for the first time. Because he had proven himself to be one of the brightest students in physics, Lee won a Chinese government scholarship to go to the United States in 1946. He opted for Chicago as his destination, and it was a lucky choice: there, in fact, under the supervision of Enrico Fermi, he obtained his doctorate in 1950, with a nuclear physics research that focused on white dwarfs. During his university years, he had spent a lot of time with Yang discussing physics and many other things. The two were complementary: if Yang put his unquestionable mathematical competence into play, Lee used his great physical intuition to weigh in. However, things were never completely clear and, in their discussions, the role of the two would often be reversed, just like in the symbol par excellence of Eastern philosophy, the Yin-Yang: there is a little bit of Yang in the Yin, and some Yin is always present in the Yang. After his doctorate, Lee spent about a year at the University of Berkeley, and he was then given the opportunity to go to Princeton, where he arrived in 1951, to find his old friend Frank Yang. In addition to nuclear physics, they both had a passion for statistical mechanics: Yang had already obtained a very relevant result on the Ising model, while Lee still had vivid lessons in his head that he had followed on the condensation of gases, held in Chicago by a famous physicist couple, Joe and Maria Mayer. It was therefore natural for the two to plunge headlong into that field, in which, in fact, they soon managed to rigorously prove many theorems on phase transitions.

Speaking about the first article that they wrote together, Lee reported, "Yang asked me to be listed as the primary author, based on the fact that he was the older of us. I found the request strange, but, in the end, I agreed, following the Chinese tradition in this way. But a quick look at the literature convinced me that this was not the way things were, so, in the following article, the authors were listed alphabetically." To general surprise, two articles were published in one of the most prestigious physics journals, the first signed by Yang and Lee, the second by Lee and Yang.

The Puzzle of the Mirror

Reverend Dodgson, also known as Lewis Carroll, was a fan of the world of mirrors and of everything that existed beyond the silver plate of their surfaces, as much of a fan as his heroine par excellence. With Alice, it all began with threats to her cat:

> And if you're not good directly ... I'll put you through into Looking-glass House. How would you like that? ... Now, if you'll only attend, Kitty, and not talk so much, I'll tell you all my ideas about Looking-glass House. First, there's the room you can see through the glass—that's just the same as our drawing room, only the things go the other way ... the books are something like our books, only the words go the wrong way; I know that, because I've held up one of our books to the glass, and then they hold up one in the other room ... How would you like to live in Looking-glass House, Kitty? I wonder if they'd give you milk in there? Perhaps Looking-glass milk isn't good to drink ... Oh, Kitty! how nice it would be if we could only get through into Looking-glass House! I'm sure it's got, oh! such beautiful things in it! Let's pretend there's a way of getting through into it, somehow, Kitty. Let's pretend the glass has got all soft like gauze, so that we can get through. Why, it's turning into a sort of mist now, I declare! It'll be easy enough to get through—

And, as we know, she actually did get through!

In 1956, physicists would have very much liked to know whether the world beyond the mirror was truly similar to ours. Until then, they had believed that, at the level of elementary particles and fundamental interactions, the world was completely symmetrical with respect to parity, i.e., symmetry by reflection. This amounted to the possibility of looking at the world through a mirror and finding it essentially similar to ours: if an event is possible—they said—its mirror image must be equally possible; an experiment conducted in a real laboratory must therefore be indistinguishable from its mirror image. If a top, viewed from above, rotates counterclockwise, its mirror image instead turns clockwise, but this does not bother us, because we could very well have spun the original top in a clockwise rather than counterclockwise motion. We can use a number equal to $+1$ or -1 to distinguish real images from mirror images: $+1$ for real images, -1 for those obtained through reflection. By inverting the mirror image twice, you return to the starting figure: this corresponds to the familiar algebraic rule

$$(-1)(-1) = 1.$$

All of the evidence accumulated up to that fateful year of 1956 had led everyone to believe that all fundamental interactions respected parity, that is, for each physical process observed, there had to be a corresponding physical process seen in the mirror.

Furthermore, in the transformation processes between the various particles, it was believed that parity was preserved, that is, the parity of the starting state should be equal to the parity of the final state.

Serious doubts, however, began to emerge on the conservation of parity from the study of a new elementary particle, the meson K, which immediately proved to be a real puzzle. At first, it was thought that there were, in fact, two distinct mesons: the first was called the theta meson, the second the tau meson, because the way in which they decayed was very different. The first decayed mainly into two pions, the second into three. However, all of the other properties of the theta and tau mesons were the same. So, why not consider them the same particle? The reason was that, if they had been the same particle, the symmetry of the physical laws under parity would have been violated! In fact, in the process in which the meson K decayed into two pions, the parity of the final state was $+1$, while, when it decayed into three pions, the final parity was instead -1. Assuming, then, the preservation of parity and the existence of only one K meson, we arrive at the paradox that it had to have $+1$ and -1 parity simultaneously, clearly a contradiction.

The catalyst for subsequent events was a lively conference held at the University of Rochester in April 1956, which saw the participation of major American theoretical physicists, including Richard Feynman, C. N. Yang and T. D. Lee. It was Feynman who called attention to the fact that the theta-tau meson paradox could be easily resolved by assuming that left-right symmetry was not respected, that is, that parity was not preserved. Those present reacted skeptically; everyone thought that it was just one of Feynman's usual provocations. An experimental physicist challenged him to accept a bet: "Would you bet a hundred dollars on the violation of parity?" Feynman immediately replied, "No, but I would certainly bet fifty dollars!" After that conference, and throughout the following summer, Lee and Yang began obsessively to think about that problem. They met in New York in early May; Yang had driven from Brookhaven National Laboratory, in the middle of Long Island, to uptown Manhattan, where Lee had an office on the Columbia University campus. By the time the clock struck 11 a.m., Yang realized that he had to move his car out of the parking lot, so they decided to go for a coffee at the White Rose Café on the corner of 125th Street and Broadway. From there, they went to lunch at the Tien Tsin Restaurant, a few meters away, and ultimately decided to return to the office. It was in those hours spent before cups of hot coffee, and then in front of a plate of Chinese dumplings, that determined what would follow: in fact, in the act of mentally reviewing a huge variety of physical processes, they realized that, while parity was preserved in the case of strong and electromagnetic interactions, there was, however, no convincing experimental evidence regarding the weak interaction! They were really excited, even elated, at this discovery. They decided to thoroughly study all of the experiments conducted up to that time on weak interactions, and subsequently proposed new ones in an article that they published in October 1956, in "Physical Review", entitled *A Question of Parity Conservation in Weak Interactions*. In it, they wrote that other experiments needed to be conducted to determine whether or not the weak interactions distinguished between the world of men and the world of mirrors.

Many experimental physicists felt prodded to accept Lee and Yang's challenge. The first was Madame Chien-Shiung Wu, of Columbia University, a Chinese scientist who had also moved to the United States. Together with a team from the National Bureau of Standards in Washington D.C., Madame Wu decided to check the directional decay properties of the cobalt-60 atoms, cooled to about absolute zero and aligned along a uniform magnetic field. In the decay process, one of the neutrons of the cobalt-60 nucleus decayed into a proton, emitting an electron and a neutrino. If the electrons had been emitted uniformly in all directions, the parity would have been preserved, while, if the emission had taken place in a preferential direction, they would have demonstrated non-conservation. Before the experiment, Wolfgang Pauli, one of the greatest theoretical physicists of all time, wrote, in a famous letter to Victor Weisskopf: "I absolutely do not believe that God can be left-handed, and I am willing to bet any amount that the experiment of Madame Wu will produce highly symmetrical results." Never was a prediction more reckless: Madame Wu's data, officially presented to Columbia University by the famous physicist Isidor Isaac Rabi on January 15, 1957, incontrovertibly showed that the electrons emitted by the nuclei had a privileged direction, and parity was not preserved! Rabi himself released a long interview that day with the New York Times in which he declared that an entire theoretical structure had been shaken by that result, and it was not known upon what roots a new one would be built. Otto Frisch, the one who discovered nuclear fission, together with Lise Meitner, on January 16, 1957, received a telegram from a friend of his, who wrote: "Parity is not preserved, here, in Princeton, they don't talk about anything else. Everyone says it is the most important result since Michelson's experiment."

In 1957, Chen-Ning Yang and Tsung-Dao Lee received the Nobel Prize in Physics, among the youngest people ever to receive this prestigious award: Lee was 30 years old, Yang 34. Unfortunately, together with the breakdown in parity, there was also a breakdown in their friendship and their scientific partnership. The reason? A quarrel over the order of their names in the publications, a topic that Yang had always been sensitive to. There was a Chinese story that told of two little boys, who, walking on the beach, had seen a light and had followed it with ever-growing enthusiasm, until they reached a palace. Upon opening the door, they were delighted by the beauty of the walls and the enormous treasure that was being kept in that building. It was the treasure of the yellow emperor, and for their discovery, they immediately became famous and greatly admired all over the world. Years later, having become old and quarrelsome, one of the two decided to engrave a gold plaque upon which was written, "Here lies the one who first saw the treasure." Seeing it, the other said, "Yes, but it was I who opened the door." The authorshop of this story is credited to T. D. Lee.

Further Reading

K.W. Ford, *A World of Elementary Particles* (Xerox College Publishing, 1963)

Zero. Someone Likes Cold

The Conquest of the Poles

The beginning of the twentieth century witnessed a great interest in the conquest of the earth's poles, the most inhospitable places on our planet. In the early years of the century, there was, in fact, a rapid succession of adventurous enterprises, often with dramatic results, sometimes animated only by the taste of challenge and risk, by the desire for glory or to excel: these enterprises conquered the front pages of newspapers, keeping their readers in suspense, making the names of the explorers legendary, and attracting the attention of Chancelleries and the interest of cartographers. In the eyes of most, it was obvious that these were challenges that sought to achieve the impossible. If the conquest of the North Pole was marked by the great rivalry between two famous American explorers, Robert E. Peary and Frederick Cook, that of the South Pole instead saw a challenge between the rough Norwegian Roald Amundsen and the British Navy captain Robert Falcon Scott. Amundsen sailed from Christiania, Norway, in August 1910, expressing his intention to reach Madeira, an archipelago in the Atlantic Ocean, but, oddly enough, he had packed numerous sledges, a hundred Greenlandic dogs and many provisions that seemed contradictory to his stated mission. After a month, he declared his real goal to the crew: to head for Antarctica and conquer the South Pole. Scott, on the other hand, sailed from Cardiff in June of the same year, and only in October, during his stop in Melbourne, did he learn, to his great amazement, that Amundsen was also headed for the South Pole! It was going to be a challenge with no holds barred …

The end of this story is tragically well known. Amundsen was the first to reach the shores of the Antarctic ice sheet and, from there, set off to conquer the South Pole: on October 20, 1911, he left his base in the Bay of Whales, on the south bank of the Antarctic continent, together with four companions, four sleds and a pack of about fifty dogs. Strengthened by the great training that they had undergone on the snow-capped mountains of Norway, the team reached the South Pole in just two months, on December 14, 1911, and returned just as easily, although they ended up having to kill and eat the dogs that they had brought with them. Scott instead

© Springer Nature Switzerland AG 2020
G. Mussardo, *The ABC's of Science*,
https://doi.org/10.1007/978-3-030-55169-8_26

left his base camp near the McMurdo Canal on November 1, 1911 with twenty-four dogs, twelve sledges, two motorized sledges, and ten ponies from Manchuria: the motorized sledges broke immediately, and the ponies had to be abandoned as well, since they had a hard time moving on ice, and it was only after a grueling march of fifty days that the British expedition arrived at the South Pole, in January 1912, to an unpleasant surprise: in the white of the Arctic pack, there were four Norwegian flags, well planted in the ice, and a tent in which two letters had been left, dated December 14, 1911, one addressed to Scott and the other to the King of England, in which Amundsen announced his conquest of the South Pole! Disconcerted and with morale in pieces, half frozen and running out of food, the British explorers embarked on the return journey, dying one after the other of cold and hunger. The last to die was Robert Falcon Scott, who, on the last page of his diary, wrote: "Had we survived, I would have had a story to tell you about the courage, the resistance and the courage of my companions that would have moved the heart of every Briton."

While the great explorers were busy with their packs of dogs in that unbridled competition to reach the coldest points on Earth, in the European laboratories, there was an equally frantic challenge, this time of a scientific nature, to reach the coldest temperature in the Universe: absolute zero.

James Dewar, the Cold Wizard

This scientific enterprise was also extraordinary because, as in the case of the famous Achilles paradox with the turtle, every time the finish line seemed within reach, in reality, after a while, it would move further, further and further away. The very existence of an absolute zero, a starting point for temperatures, appeared, to the experimental physicists of the time, as a mysterious phenomenon, an extreme frontier of the laws of Nature. However, it was a frontier so fascinating as to push scientists from half of Europe to start a breathless race to be the first to reach the most extreme cold, the most glacial goal imaginable. Like that of the polar explorers, this is a story of successes, defeats, great rivalries and deep states of despair, because there was a very ambitious goal at stake: for the winner, the glory and the opportunity to win the Nobel Prize, for the loser, the prospect of ending up in a forgotten corner of the history of science.

When the first men entered the ice of the Antarctic, they felt the chill of the coldest cold on the globe on their skin, about 80° below zero centigrade, but this was nothing compared to the ultimate limit of frost: on the Celsius scale, absolute zero is set at about −273° with respect to the melting temperature of ice. Among the first scientists willing to reach this hypothetical ultimate limit was James Dewar, professor at the Royal Institution of London. One evening in 1881, during one of his famous public lectures, he had made the stakes clear: "If we could get closer to absolute zero, we would open a new horizon for science and, at the same time, a new window on the states of matter." James Dewar was a very skilled physicist and an equally ambitious man, a pragmatic Scotsman and an expert storyteller, capable of showing

both his colleagues and ordinary people the wonders of cold and the secrets that it revealed. The rare fusion of experimental ability and stage presence, supported by a profound scientific intelligence, made him a very fascinating personality. A large canvas exhibited at the Royal Institution depicts him in an academic gown, with his perfectly groomed beard and his penetrating gaze, while he grapples with the demonstration of the properties of hydrogen in front of a large audience. His lectures on Friday evening, in the tradition famously inaugurated by Faraday, were all the more extraordinary as they were real social occasions, during which the public could witness the drama of scientific progress. On one of these evenings, after repeatedly bouncing a rubber ball on the table, he immersed it in a steaming ampoule full of liquid oxygen and, after extracting it with forceps, crushed it into a thousand pieces with the blow of a hammer, causing general amazement: Dewar wanted to show that, by changing the temperature, the physical properties of the material changed as well and, at the temperature of the liquid oxygen, the original elastic ball had turned, before the astonished eyes of the spectators, into a solid body, hard as stone.

These phantasmagorical experiments certainly helped nineteenth-century science to gain the attention of the general public and, despite being, at times, a little mystifying, they gave the impression that, in the laboratories, those strange scientists were doing something really great, almost magical. Some of these researches had also produced interesting practical effects, for example, the invention of the thermos, achieved by Dewar in the course of his studies on cryogenics: two glass containers, placed one inside of the other and separated by a cavity with a silver surface, in which there is very little air. Now, glass is already a very bad conductor of heat; the lack of air in the interspace thus prevents heat transfer and the silvering of the walls further reduces dissipation. Put simply, the Dewar jar is ideal for keeping the liquids put inside of it at a constant temperature for a long time, a fact that was first noticed by the glassblower from whom Dewar had commissioned the ampoule, upon finding that the milk he had put in the container for his son the night before was still warm in the morning.

Box: Superconductivity and Superfluidity

At temperatures close to absolute zero, some substances exhibit amazing behaviors: their electrical resistance can completely vanish, making them "superconducting"! This has no explanation in classical physics, but the origin of this effect lies in the laws of quantum mechanics that regulate systems with a huge number of particles: it is necessary to bring up the Bose-Einstein statistics and the phenomenon of condensation to which it gives rise when approaching absolute zero. In addition to having zero electrical resistance, superconductors have other singular properties: for example, subject to an external magnetic field, they are able to expel it completely from their inside. It was Kamerlingh Onnes himself who discovered, in 1911, the phenomenon of superconductivity by studying the electrical properties of mercury once it had been immersed in liquid helium: below a certain temperature close to absolute zero, Onnes observed that the resistance of the mercury completely vanished. The same strange property was noticed years later in many other materials, such as lead, cadmium, aluminum, etc. Recently, it has been discovered that some ceramic materials also become superconductors at relatively high temperatures, paving the way for important technological applications, such as the construction of magnet coils using superconducting cables (very useful in medical machines for magnetic resonance imaging). In elementary particle physics, superconductors find application in large detectors installed at CERN around accelerators. The super sensitivity standards of superconductors also make them ideal for precision measurements of very small magnetic fields, fundamental over a vast range of different sectors, from medicine to geology.

Liquid helium also has the ability to flow without any friction: this phenomenon of "superfluidity" was also discovered by Kamerlingh Onnes in 1911, although the scientist was unable to fully understand its implications. We had to wait until 1937, and Pyotr Kapitsa's studies in Moscow, to fully appreciate the surprising characteristics of a quantum nature. Helium is the only substance that remains liquid at ordinary pressure and close to absolute zero, since every other substance passes into the solid state if cooled at such low temperatures. According to thermodynamics, temperature is a measure of the velocity of the atoms of which a substance is made, and therefore it is expected that, going towards absolute zero, the atomic motion will stop and the material will become solid. The fact that helium remains liquid even at absolute zero is an initial indication of the exceptional character of this element. It is able easily to cross very thin capillaries, an achievement that represents an insurmountable obstacle for normal fluids, endowed with viscosity. Lately, super-fluidity phenomena have also been discovered in the field of cold atoms, a result that has given rise to an explosion of theoretical and experimental studies. A demonstration of the great importance of this research is confirmed by the fact that, in just fifty years, as many as 19 Nobel Prizes have been awarded for works related to these themes.

The life of certain men is summed up by a singular obsession, and for James Dewar, this obsession was cold. Since he was a boy, he had loved to skate on the frozen surface of the Scottish lakes. He claimed that the most formative experience of his life had been a fall into the freezing water of a pond: he was pulled out immediately, but, once home, he had developed a very high fever that forced him to stay in bed for the following six months, long wavering between life and death. Despite having escaped the worst, his fingers remained locked. To limber them up, he asked the village carpenter to teach him to build a violin: the long hours spent honing the maple of the chest and carving the pegs and the nut to wrap the strings gave him all of the manual skills necessary for the construction, years later, of sophisticated laboratory instruments. He never fully recovered the movement of his hands, but, in return, he developed an infatuation with cold that was destined to condition his scientific life forever.

James Dewar's dream was to take up the baton left by the most important scientist of the Royal Institution, Michael Faraday. Many years earlier, Faraday had shown how a gas such as chlorine could be liquefied by applying a sufficiently large amount of pressure, also noting that, when the liquid evaporated, the temperature dropped drastically. Faraday became intrigued and decided to see if all gases could liquefy through the use of pressure; to his enormous surprise, however, he noticed that some of them—which he called the "permanent gases"—did not liquefy, even when enormous pressure was applied. Discouraged by his failures, Faraday abandoned this line of research: in the following thirty years, no one else was able to go below the record temperature that he had obtained, which was around −130°. Absolute zero therefore remained a very elusive goal, a real unknown land, a territory difficult to reach. This was the great challenge that faced the scientists of the nineteenth century.

It was only in 1873 that Johannes Diderik van der Waals, a unique Dutch theoretical physicist, was able to explain why, until then, it had not been possible to liquefy gases such as hydrogen, oxygen and nitrogen. In fact, thanks to a very acute theoretical idea that allowed him to estimate the size of the molecules and their reciprocal forces, he showed that, to liquefy these gases, it was necessary not only to apply strong pressure, but also to go below a critical temperature, thus indicating the key to transforming the so-called permanent gases into liquids for the first time. The first to capitulate was oxygen (at −218°), followed immediately afterwards by nitrogen (at −200°), and finally by hydrogen (at −250°). It was Dewar who liquefied the latter on May 10, 1898, producing a small amount of liquid, which he observed quietly boiling in a vacuum-sealed ampoule: another stage had been reached in the conquest of absolute zero. Dewar stated that the next step was the liquefaction of helium, a process that, according to his estimates, should have taken place at very few degrees above absolute zero: this gas, very abundant in the Sun and in the rest of the stars, but very scarce on Earth, seemed reluctant to any change of state, and succeeding in liquifying it would have been one of humankind's most important successes, a real triumph of science. Dewar was determined to be the first to come that close to absolute zero, but, unfortunately, he was not alone in this race.

Heike Kamerlingh Onnes, Knowledge Through Measurement

There were several groups involved in this race—both in Poland and France—but the rival that Dewar feared most was a brilliant Dutch physicist, Heike Kamerlingh Onnes, about ten years younger than him. Onnes had become a professor of physics at the University of Leiden in 1882, at just 29 years of age, after an exceptional academic career. In the inaugural speech of his course, he focused on the importance of quantitative research; in particular, he hoped that the observations made in the laboratory would be carried out with the same precision and accuracy with which astronomers measured the positions of the stars of the celestial vault. On that occasion, he proposed posting his motto—'*Door meten tot weten*,' 'Knowledge through Measurement'—on the door of every laboratory. In addition to a solid theoretical background, Onnes possessed an impeccable control of scientific instruments, combined with a remarkable ability in the creation of experiments and a real love for the numbers that came out questioning nature.

Dewar knew well that the Dutch physicist had a radically different approach to science than he did. The two, in fact, could not have been more dissimilar: if Dewar was obsessed with secrecy, so much so as to refuse to reveal to anyone the protocol and tools that he used for his experiments, Onnes loved to make his discoveries public, publishing the details in the major scientific journals of the time, and always keeping his laboratory notebook in order, full of notes, observations, and detailed comments. Dewar used to treat his collaborators badly, often engaging in furious quarrels that put an end to their professional relationship. Onnes was, instead, diplomatic, very careful and patient: he was extremely attentive to his technicians and laboratory assistants, who, thanks to him, felt that they were part of a great scientific enterprise, and therefore tolerated his exhausting rhythms of work without complaint. It will be enough to quote an anecdote: when Onnes died, at some point, the horse pulling the cart with his coffin began to accelerate, forcing all of those participating in the ceremony to increase their pace. "He forces us to run even when he's dead," was the ironic comment from his foreman.

Onnes had an organizational spirit that was out of the ordinary, with the mentality of an industrial captain. His laboratory, in fact, operated like a small factory, where expert glassmakers, chemists and instrument builders were employed, affectionately known by him as the "blue boys", since they worked constantly immersed in the blue light of the Bunsen burners. He was, perhaps, the first scientist to understand the complexity linked to scientific enterprises, which require large quantities of materials, sophisticated instrumentation, research funds, and dedication and competence from the staff. It was in this spirit that he prepared himself for the demanding challenge of liquefying helium and traveling into the unknown land of temperatures close to absolute zero.

First of all, Onnes understood that it was crucial to have a sufficient quantity of helium of adequate purity. This gas was extracted from monazite sand, and therefore the scientist obtained considerable quantities of this material over time. After a few weeks, thanks to the tireless work of his technicians and chemical assistants, he managed to dispose of about 200 L of gaseous helium. It was now necessary to make use of the "cascade method", a procedure that had proved extremely effective in the field of low temperatures. It was practically identical to the technique used by climbers to reach the most inaccessible peaks, that is, to make use of a series of base camps set at different heights from which to launch the assault on higher heights. In the field of cryogenics, this method consisted of using the liquefaction temperature of a gas as a starting point for the liquefaction of another gas at a lower temperature, and so on, so as to gradually descend towards increasingly freezing temperatures. The cascade method required putting the ampoules containing the various gases into contact: this was why the blast furnaces and glass blowers at the Leiden laboratory had been working at full speed for days, to produce test tubes, ampoules and valves strong enough to withstand such low temperatures and at such great pressure variations. All of this was ready by early June 1908.

While preparations for the great experiment were going on in Leiden, in London, James Dewar was struggling with a serious problem: two steps away from the Royal Institution were the laboratories of William Ramsay, the chemist who was responsible for the discovery of almost all noble gases, such as argon, neon, kripton, xenon and helium. Dewar could have easily taken advantage of this fortunate proximity to acquire huge quantities of liquid helium of the highest quality, except that he had recently quarreled mightily with Ramsay, so much so that the two were no longer speaking. It had all started a few months earlier at the Royal Society over questions of priority in the liquefaction of hydrogen: Ramsay claimed that the first to achieve it was the Polish physicist Olszewski; Dewar was highly irritated by this statement, and the tone of the discussion quickly turned, almost degenerating into insults. Since Dewar could no longer count on Ramsay's help to obtain appreciable quantities of helium, he decided to build a large machine himself to isolate and purify helium from neon, an instrument that he never had any luck operating, because the valves were systematically blocked with frozen neon. The coup de grace was a laboratory accident caused by a technician, who, in carelessly turning a valve, had dispersed all of the helium that Dewar had managed to accumulate up to then into the air.

Arrival and Departure Points

Kamerlingh Onnes liquefied helium on July 9, 1908, on a wet and windy day. He got up at 4 that morning so as to get to the lab early. The cascade process went on for hours, and it was only towards the late afternoon that the first drops of a new liquid began to be noticed, at an estimated temperature of $-269°$, just $4°$ above absolute zero! At that precise moment, Leiden became the coldest place on earth, at least as cold as the faraway sidereal spaces. That experiment not only brought us to the threshold of absolute zero for the first time, it also opened up the doors of a new, mysterious and fascinating world, a world where substances lose their electrical resistance and fluids flow without friction, a world dominated by truly extravagant effects. And if the name of Kamerlingh Onnes is now written in large letters in the Pantheon of science, that of James Dewar has unfortunately been lost in the folds of history, in the oblivion of time and memory.

Further Reading

K. Mendelssohn, *The Quest for Absolute Zero* (Taylor & Francis, 1977)